# ANTARCTIC ATLAS

*New Maps and Graphics That Tell the Story of a Continent*

.

PARTICULAR BOOKS

UK | USA | Canada | Ireland | Australia
India | New Zealand | South Africa

Particular Books is part of the Penguin Random House group of companies whose addresses can be found at:
global.penguinrandomhouse.com.

First published 2020
001

Printed and bound in Germany by aprinta druck GmbH

A CIP catalogue record for this book is available from the British Library

ISBN: 978-1-846-14933-7

www.greenpenguin.co.uk

MIX
Paper from
responsible sources
FSC
www.fsc.org    FSC® C018179

Penguin Random House is committed to a
sustainable future for our business, our readers
and our planet. This book is made from Forest
Stewardship Council® certified paper.

# ANTARCTIC ATLAS

*New Maps and Graphics That Tell the Story of a Continent*

Peter Fretwell

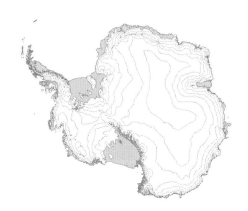

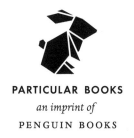

**PARTICULAR BOOKS**

*an imprint of*

PENGUIN BOOKS

# CONTENTS

Introduction *7*

*Chapter 1* **THE GEOGRAPHY OF ANTARCTICA**

1. The white continent *14*
2. The seven poles *16*
3. The meaning of Antarctica *18*
4. A continent of many lands *20*
5. Size matters *22*
6. The shelf and the sheet *24*
7. The night, the light and the half-light *26*

*Chapter 2* **ICE**

8. The ice beneath your feet *30*
9. Ice flow *32*
10. A changing world *34*
11. Drowning coasts *36*
12. The anatomy of an ice sheet *38*
13. Draining the depths *40*
14. Antarctic time machine *42*
15. Shrinking shelves *44*

*Chapter 3* **LAND**

16. Rocks below the ice *48*
17. The hidden world *50*
18. The making of Antarctica *52*
19. Volcanoes *54*
20. The quaking sea *58*
21. The driest place on earth *60*
22. Alien invasion *62*
23. Majestic mountains *64*
24. The wolf at the end of the world *68*

*Chapter 4* **ATMOSPHERE**

25. Don't forget your thermals *72*
26. The hole at the bottom of the world *74*
27. A future in our hands *78*
28. Stormy weather *80*
29. Polar vortex *82*

*Chapter 5* **SEA**

30. The Southern Ocean *86*
31. Islands in the stream *88*
32. Ocean currents *92*
33. Ocean eddies *94*
34. The greatest change on Earth *96*
35. The engine of the ocean *98*
36. The life of a berg *100*
37. The green ocean *102*
38. Earth's lungs *104*

*Chapter 6* **WILDLIFE**

39. Keystone krill *108*
40. The realm of the emperor *110*
41. An ocean of penguins *112*
42. International seal travels *114*
43. The blood-red sea *116*
44. The great wanderers *118*
45. The richest place on earth *120*

*Chapter 7* **PEOPLE**

46. Going south *124*
47. Who lives there? *126*
48. Sweet home Antarctica *128*
49. Pieces of pie *132*
50. Who owns it? *134*
51. Mac Town *136*
52. Mobile home *138*
53. International antics *142*
54. Antarctic skies *144*
55. Exploiting the ocean *146*
56. Tourist hub *148*

*Chapter 8* **EXPLORATION**

57. To find a continent *152*
58. The heroic age *154*
59. The greatest escape *156*
60. The race that never was *160*
61. Get on your knees and pray *164*
62. The home of the blizzard *168*
63. Exploring from above *172*
64. Postwar power play *174*
65. The scientific age *176*
66. The satellite age *178*
67. A most historic place *180*
68. Traces of the past *182*

*Chapter 9* **FUTURE**

69. Looking ahead *180*
70. Antarctica 10,000:
    The distant future *188*

Glossary *193*

Acknowledgements *197*

References, data sources and
further reading *199*

Index *203*

Images *207*

# *Introduction*

**IT'S FEBRUARY 2018**, and I am sitting in a bright orange tent, pitched on a glacier 12 kilometres wide. This frozen river of ice, the Sibelius Glacier, is located in the centre of Alexander Island, Antarctica's largest island and the second largest uninhabited island on earth. Very few people have been here before, and tomorrow my field assistant and I will ski south to explore and investigate part of a small range of hills that no one has yet visited. A place where no human feet have yet trodden. There are fewer and fewer places on earth where this claim can be made.

**I LOOK OUT** of the tent flap, across the featureless white river of ice to the soaring bulk of the mountain to the east. Its name is Mount Stephenson, and, like almost every other peak in the area, it has never been climbed. The mountain's western face rises almost vertically 3 kilometres from the brilliant white, pristine glacier to loom over the local hills. The scale and grandeur of the natural environment render me speechless, and I stare out at the scene, listening to the murmur of the wind, which, apart from my own heartbeat, is the only sound in this immense frozen world. Our tent is insignificant in a vast white landscape. Here, away from the frenetic bustle of civilization, people are irrelevant. But I know things are not that simple.

The snow and ice may look immortal and pristine, but this is a landscape on the brink.

Ten months later, and I am back at my desk in Cambridge. I am looking at a newly downloaded satellite image of the Wilkins Ice Shelf, a large, floating mass of ice into which the Sibelius Glacier flows. The ice shelf is fracturing and disintegrating. It is a process that started almost a decade ago. This ice shelf, like most of the others on the Antarctic Peninsula, is dying. The oceanographers tell me that it is due to warmer seawater penetrating beneath the floating shelf and melting it from below. Just one more sign of global warming in this fragile region. The Antarctic Peninsula is an area that has warmed by over 4°C in the last fifty years. Once the ice shelf has gone, the glaciers behind it, including Sibelius, will accelerate, flowing faster into the sea. Their ice surface will fall, more rock outcrop will become exposed and, over time, their great fields of ice will disappear. And so a landscape largely unseen by human eyes may disappear because of human-made climate change.

And it is not just the ice. The biology and wildlife, from phytoplankton to penguins, are threatened by change. Only half of all emperor penguin colonies have ever been visited by humans, and yet in a hundred years' time less than half of the colonies that exist today will still be here. This is not an arbitrary concept. Antarctica might be remote and otherworldly, stuck away at the bottom of the planet, seen only by rich tourists and on wildlife documentaries, but it is vital. The oceans around the continent control the world's weather systems and soak up half of the extra atmospheric carbon that humans produce. If it melts, the land carries enough ice to swamp the world's coastal cities. Its fragile environment acts as a bell warning of the environmental consequences that could affect other parts of the world. Far from being abstract, Antarctica has direct relevance to us all.

This is one of the reasons I have written this book. On one hand, I wanted to share my wonder at Antarctica, its awesome beauty and harsh magic. But on the other, I also wanted to try to explain the importance of the white continent and to tell the story of change in this, the most remote of all places. I hope that I can convey some of my enthusiasm and love of Earth's last great wilderness as well as my concern for that environment.

This is a book of maps. Each map is new, and many of the concepts mapped have never been presented in this way before. The map is a wonderful medium for conveying information. I have always felt that, if a picture is worth a thousand words, a map can tell a whole story. I have been making maps professionally for twenty years, and I love them as a way of communicating abstract ideas. I know that getting the balance right between artistry and information, detail and simplicity, is a delicate compromise. Maps can be things of beauty, but they must also convey a clear message or story to the viewer. The best maps do both:

we understand them intuitively and they often delight and entertain us by revealing hidden detail on closer inspection. This book uses maps to highlight and invigorate what could otherwise become dense and complex scientific messages. Each map tells its own story, and the book is designed for each page to stand alone so that you can dip in or out or browse specific sections.

During my research I spoke with many of the world's experts about what is happening in Antarctica: glaciologists, oceanographers, geographers, historians, logistics experts, modellers, ecologists, geologists and geophysicists. I am lucky to be in a virtually unique position. I have worked at the British Antarctic Survey (BAS) for eighteen years, and over much of that time my job has been to make maps. I calculate that I have made more than two thousand – for scientific papers, posters, presentations, analysis and reports. As well as maps for the BAS, I have made maps for the Intergovernmental Panel on Climate Change, the Scientific Committee on Antarctic Research and the Foreign Office and for many other books and publications. I have also made a number of published wall maps, but over the years, as the analysis becomes more complex and the techniques I use more original, my job has changed from cartographer to scientist. Over the last decade I have authored or co-authored fifty research papers, and I am lucky enough to have collaborated on many projects with leading scientists across disciplines. This has given me

access to prominent scientists around the world who work on Antarctic studies, many of whom have helped with the research for this book.

I have spent four field seasons in Antarctica, sometimes on a research station, sometimes on a ship and sometimes in a tent. Being an Antarctic scientist is a privilege, one that often we don't fully appreciate. We view places few others will ever see and experience environments most people will never encounter. It is almost always physically demanding, sometimes dangerous and often disappointing. Antarctica can be cruel, and the severity of the environment destroys the hopes and ambitions of many researchers. But when we get back and have time to contemplate, almost everyone who has been out there finds that the Antarctic has touched them deeply, leaving an indelible mark on their soul.

That is my overriding feeling about Antarctica. It is a place apart, like nowhere else on Earth. It holds an almost mythical fascination, and in a world that is becoming ever more connected, smaller and more civilized, the thought that this untouched white wilderness still exists renders it even more alien and otherworldly. How long the continent will stay that way and what its future holds are still to be decided. One of the other lessons that I have learned in the making of this book is that, although change to our environment is inevitable, how much change and what future we choose is up to us. It is still not too late to make that choice.

**PETER FRETWELL**, *December 2019*

**THE BRITISH ANTARCTIC SURVEY** (BAS), an institute of the Natural Environment Research Council, delivers and enables world-leading interdisciplinary research in the polar regions. Its skilled science and support staff, based in Cambridge, Antarctica and the Arctic, work together to deliver research that uses the polar regions to advance our understanding of Earth and our impact on it.

Through its extensive logistic capability and knowhow, BAS facilitates access for the British and international science community to the UK polar research operation. Numerous national and international collaborations, combined with an excellent infrastructure, help sustain a world-leading position for the UK in Antarctic affairs. The UK is one of over thirty countries operating scientific research facilities in Antarctica.

BAS operates two year-round and three summer-only research stations in Antarctica, as well as two stations on South Georgia in the Southern Ocean. It maintains the stations and supports many field parties with the help of two polar ships and a fleet of five specialist aircraft. Over the past fifty years, BAS has been responsible for many scientific breakthroughs, including the discovery of the ozone hole, and has led major interdisciplinary projects to extract ice cores and to understand, map and model the Antarctic continent.

The vision of BAS is to be a world-leading centre for polar science and polar operations, addressing issues of global importance and helping society adapt to a changing world.

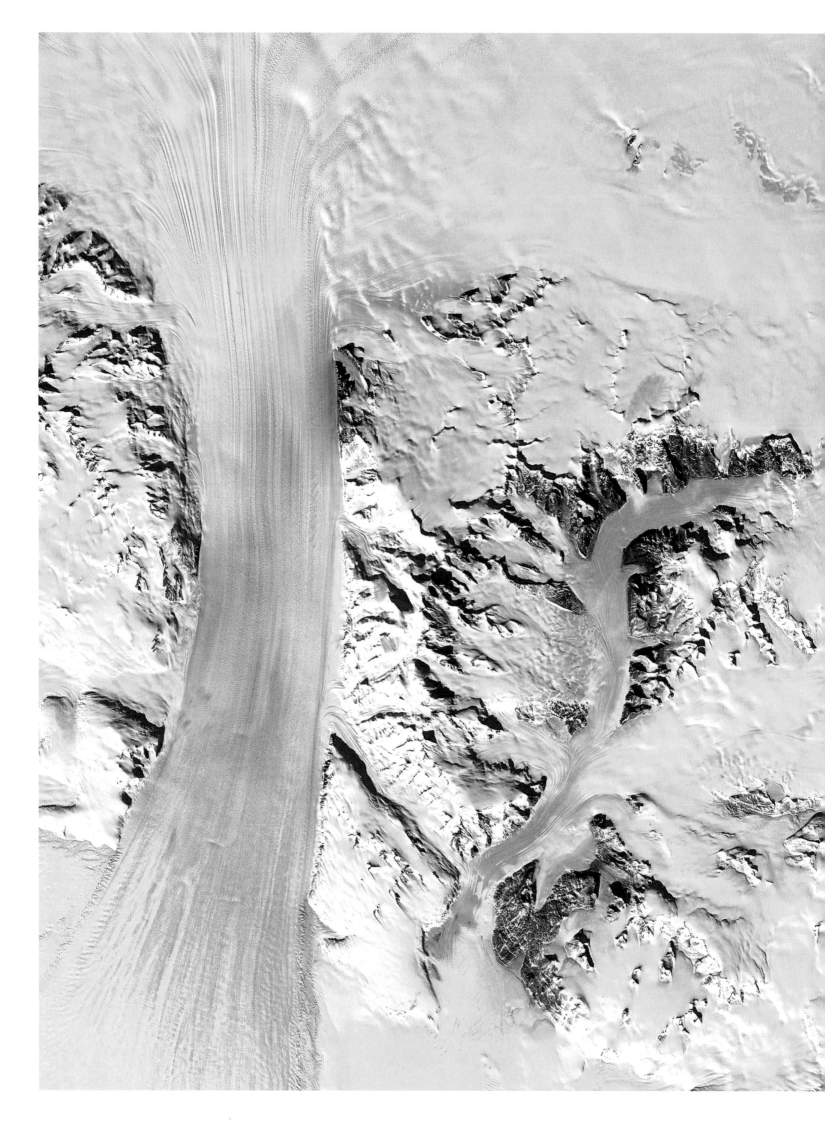

# CHAPTER 1

# The Geography of Antarctica

1. The white continent  *14*

2. The seven poles  *16*

3. The meaning of Antarctica  *18*

4. A continent of many lands  *20*

5. Size matters  *22*

6. The shelf and the sheet  *24*

7. The night, the light and the half-light  *26*

# 1. *The white continent*

**THIS MAP SHOWS** the general shape and geography of Antarctica. The continent is often split into three parts: the Antarctic Peninsula, West Antarctica and East Antarctica.

The rocky arm of the Antarctic Peninsula stretches 1,000 kilometres northwards from the main continental land mass. Geologically, it is an ancient part of the Andes Mountains in South America, and its steep sides fall precipitously into the sea, with many deep fjords and islands.

West and East Antarctica are dominated by vast ice sheets, immense, featureless domes of ice several kilometres thick. They are separated by the Transantarctic Mountains, the great mountain chain that dissects the continent. East Antarctica is larger, and its ice sheets are higher and thicker. These ice sheets rise to a series of ice domes. The highest, in the centre of the continent, is over 4,000 metres above sea level. West Antarctica is smaller and lower and lacks the great ice sheets of East Antarctica.

Most of the continent is surrounded by ice shelves – these are flat platforms of floating ice formed when glaciers flow into the ocean, but, rather than breaking off to form an iceberg, the floating ice merges with other glaciers to create a flat expanse of floating ice. These shelves can be hundreds of metres thick and can extend for hundreds of kilometres away from the coast. On either side of the continent, between East and West Antarctica, major indentations in the coastline give rise to two massive shelves – the Ross Ice Shelf and the Ronne-Filchner. The Ross Ice Shelf is larger than Spain and the Ronne-Filchner is only a little smaller.

0°

30°E

60°S

60°E

Dronning Maud Land

Enderby Land

65°S

Coats Land

*dell Sea*

DOME F

Kemp Land

70°S

Mac Robertson
Land

*onne-Filchner
Ice Shelf*

75°S

*Amery Ice Shelf*

*Prydz Bay*

**EAST ANTARCTICA**

80°S

Queen Elizabeth
Land

Princess
Elizabeth
Land

DOME A
4,000

**SOUTH POLE**

Wilhelm II Land

90°E

Ellsworth Mountains

**TRANSANTARCTIC MOUNTAINS**

3,500

Queen Mary Land

3,000

2,500

Marie Byrd
Land

2,000

DOME C

Wilkes Land

1,000

*Ross Ice Shelf*

Terre Adélie

120°E

*Ross Sea*

Victoria Land

George V Land

Oates Land

0   250   500   1000

KILOMETRES

60°W

150°E

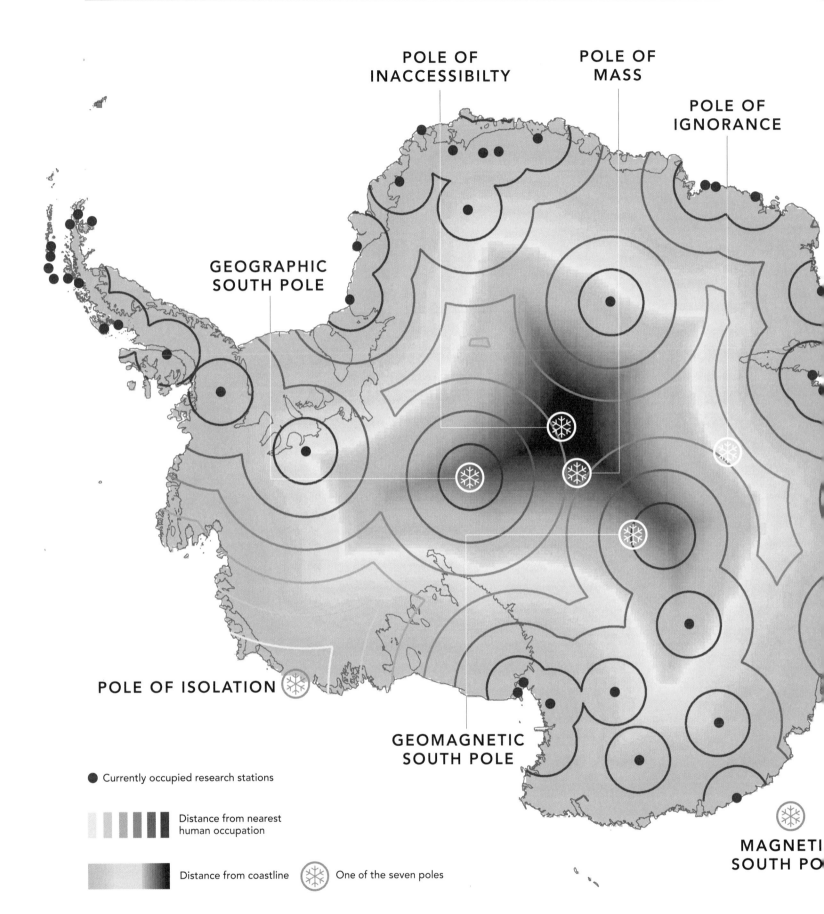

POLE OF
INACCESSIBILTY

POLE OF
MASS

POLE OF
IGNORANCE

GEOGRAPHIC
SOUTH POLE

POLE OF ISOLATION

GEOMAGNETIC
SOUTH POLE

MAGNETI
SOUTH PO

● Currently occupied research stations

Distance from nearest
human occupation

Distance from coastline    One of the seven poles

# 2. *The seven poles*

EVERYONE KNOWS THAT the South Pole is in Antarctica, but this, the Geographic South Pole, (0°E, 90°S) is only one of several poles on the continent. As well as this geographic pole, there two types of magnetic pole. The first is the true Magnetic South Pole, the point where a southward-facing compass arrow points vertically down. This position wanders between 10 and 15 kilometres each year, owing to the movement of molten rock deep in the Earth's interior, and is not symmetrical between the North and South Poles. Then there's the Geomagnetic South Pole, a fixed, symmetrical approximation of the true position, that geographers and navigators find easier to work with. The magnetic pole is off the coast of East Antarctica, presently at 136.1 east, 64.2 south, while the geomagnetic pole is deep in the continental interior, high on the East Antarctic ice sheet, at 107.0 degrees east, 80.5 degrees south.

In addition to these three classical poles, there are a number of others. The first is the Pole of Isolation; the point in Antarctica furthest from permanent human occupation. This spot, on the Marie Byrd Land coast at 138 degrees west, 75 degrees south, was once the site of a Russian research station, but this has been abandoned since 1980. If you were standing there today, you would be over 1,400 kilometres from the nearest human habitation, in possibly the most isolated location in the world.

The next most well-known pole is the Pole of Inaccessibility, the place furthest from the coast. This is at 64.7 degrees east, 83.9 degrees south, high on the East Antarctic plateau, 1,580 kilometres from the open sea.

Next is the Pole of Ignorance, a term coined for the area we know least about. In the context of Antarctica, this is the poorest-mapped part of the Antarctic subglacial bed, and it is also in East Antarctica, in Princess Elizabeth Land, at 85.5 degrees east, 75 degrees south.

Finally, there is the Pole of Mass. This is where the centre of gravity of the vast Antarctic ice sheet rests. Not surprisingly, this is close to the centre of the high East Antarctic ice sheet, at 89 degrees east, 83.5 degrees south.

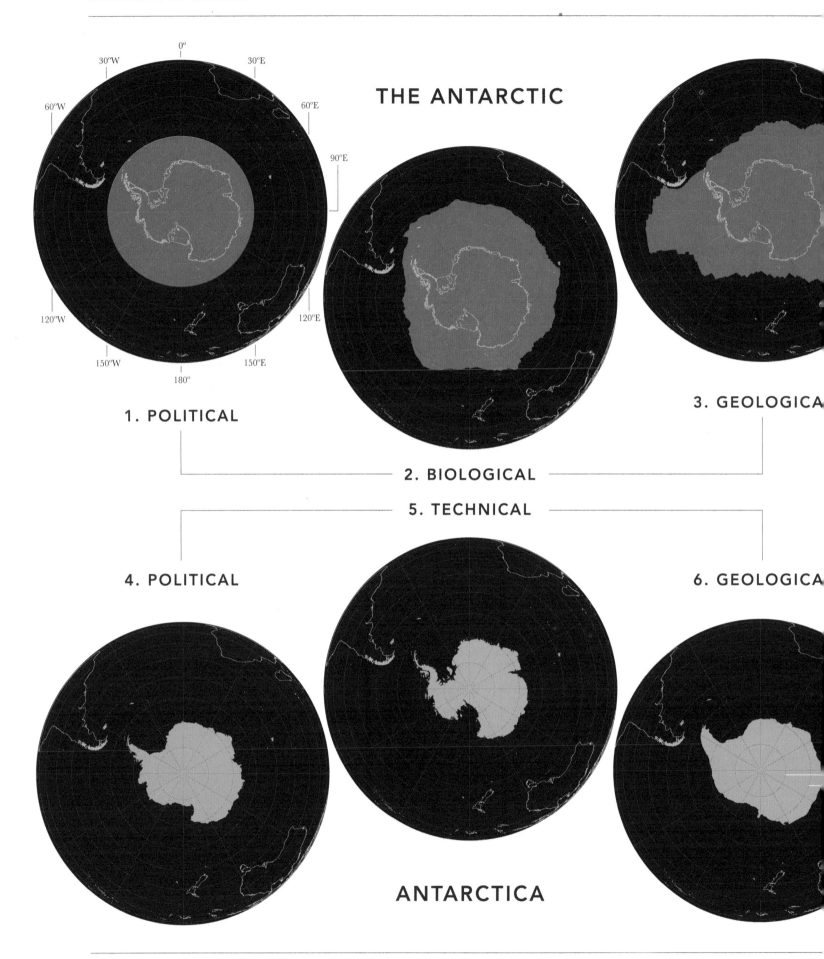

# THE ANTARCTIC

1. POLITICAL

2. BIOLOGICAL

3. GEOLOGICA

5. TECHNICAL

4. POLITICAL

6. GEOLOGICA

ANTARCTICA

# 3. The meaning of Antarctica

**ANTARCTICA IS THE** continent surrounding the South Pole, the southern end of the axis on which the Earth turns, but how do we define which bits are and which bits are not Antarctica? The term 'the Antarctic' includes the seas of the Southern Ocean, whereas 'Antarctica' refers purely to the land. But what exactly constitutes land here? The difference between this and other parts of the world is the ice shelves, which are just frozen water floating on unfrozen water. Should we count this as land? Do the offshore islands count as Antarctica and, if so, which ones? Like many areas of geography, the answer depends on who you are and what you want the answer to be.

## The Antarctic

1. *Politically*, all sea and land below 60 degrees south. This boundary is the basis for the Antarctic Treaty, the political agreement that governs the continent.
2. *Biologically*, the area below the sharp temperature gradient in the ocean surface that separates the warm subtropical water from the cold polar waters, the 'polar front'. This definition is often used to define the Southern Ocean.
3. *Geologically*, the Antarctic Plate, the tectonic plate on which Antarctica sits.

## Antarctica

4. *Politically*, the coastline of the Antarctic continent, including all the islands, land and ice shelves below 60 degrees south. This is, technically, what the Antarctic Treaty defines Antarctica as.
5. *Technically*, the coastline of grounded ice, not including floating ice shelves. If you are of the opinion that floating ice cannot be classed as land, this would be your definition.
6. *Geologically*, the continental block inland of the Antarctic continental shelf on which the main land mass sits. This does not contain several island groups (such as the South Orkney Islands) included in the political definition.

80°S
70°S
60°S
50°S
40°S
30°S
20°S
10°S

# 4. A continent of many lands

**PALMER LAND**

**ELLSWORTH LAND**

**WHEN EXPLORERS FIRST** sighted the mainland of Antarctica, they each named the part of the coast they landed on, explored and charted, and over the years the coastline was divided up into various portions called 'lands'. As the first explorers all came by boat, they defined only the coasts, usually between prominent capes or headlands or perhaps up to a large island or ice shelf. Having no way of exploring the hinterland behind the coastal strip, the inland extent of each land was never specified, and over time we have imagined that each land extends inwards from the coast towards the pole, splitting the continent into segments, like pieces of pie.

Some of the lands, like Graham Land and Palmer Land (both on the 'Antarctic Peninsula'), are named after the explorers who first sailed down the coasts there, but later captains named lands after royalty or, occasionally, after a member of their family. Several portions of the continent that could not be reached by ship remained unnamed for many decades. The most recent land to be named was a segment south of the Ronne-Filchner Ice Shelf, between 20 and 80 degrees west, which was named Queen Elizabeth Land on the queen's Diamond Jubilee in 2012. Yet there remains one unnamed segment of Antarctica, a small portion of the continental mainland due south of the Ross Ice Shelf that encompasses part of the Transantarctic Mountains and high polar plateau.

Today, each country engaged in Antarctic research has the right to name newly discovered features, although this can lead to confusion when two countries name the same place differently, as sometimes happens. In one of its many coordinating roles, the Scientific Committee on Antarctic Research tries to coordinate place-naming efforts between nations to reduce duplication and uncertainty.

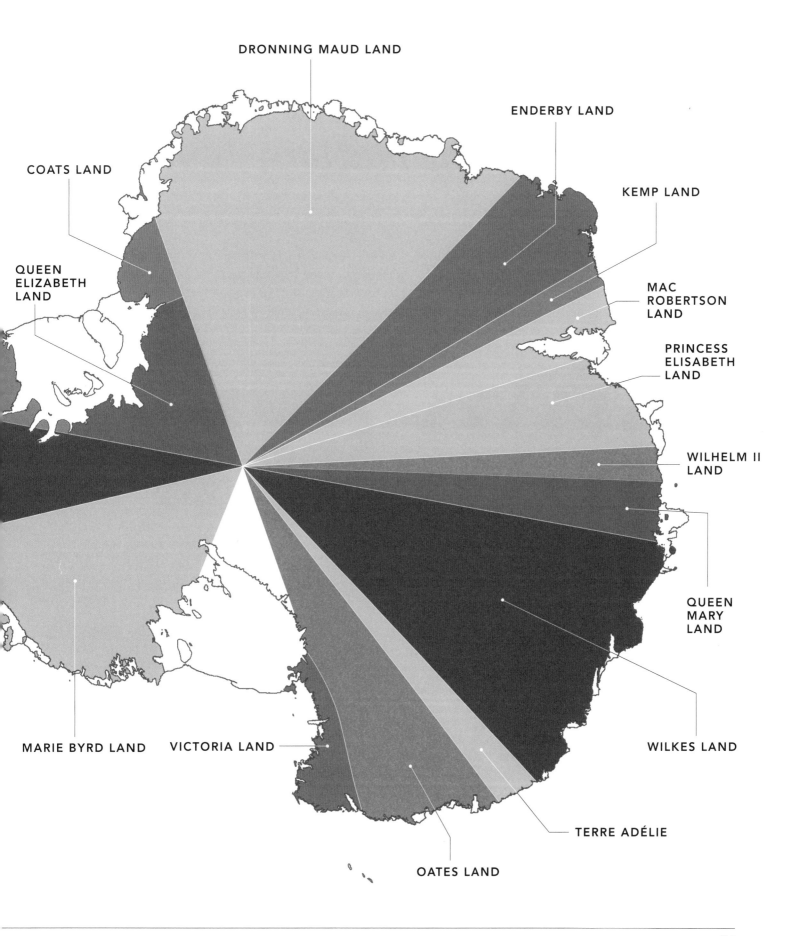

DRONNING MAUD LAND

ENDERBY LAND

COATS LAND

KEMP LAND

QUEEN
ELIZABETH
LAND

MAC
ROBERTSON
LAND

PRINCESS
ELISABETH
LAND

WILHELM II
LAND

QUEEN
MARY
LAND

MARIE BYRD LAND

VICTORIA LAND

WILKES LAND

TERRE ADÉLIE

OATES LAND

# 5. Size matters

**THE SHEER SIZE OF ANTARCTICA** is difficult to conceive. If it was a country, it would be the second-largest in the world, after Russia, and flying across it, from the tip of the Antarctic Peninsula to the far side, would involve the same distance as that between London and New York. While we usually think of Antarctica as a single entity, its sheer size means that it has many regional differences, in climate, biology and environment. These graphics attempt to convey this sense of scale, comparing the geographical area of Antarctica with several other well-known land masses.

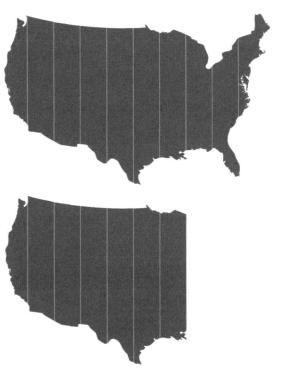

1.7 TIMES THE SIZE
OF THE US CONTIGUOUS STATES;
1.4 TIMES THE SIZE OF THE ENTIRE USA

57 TIMES THE SIZE OF THE UK

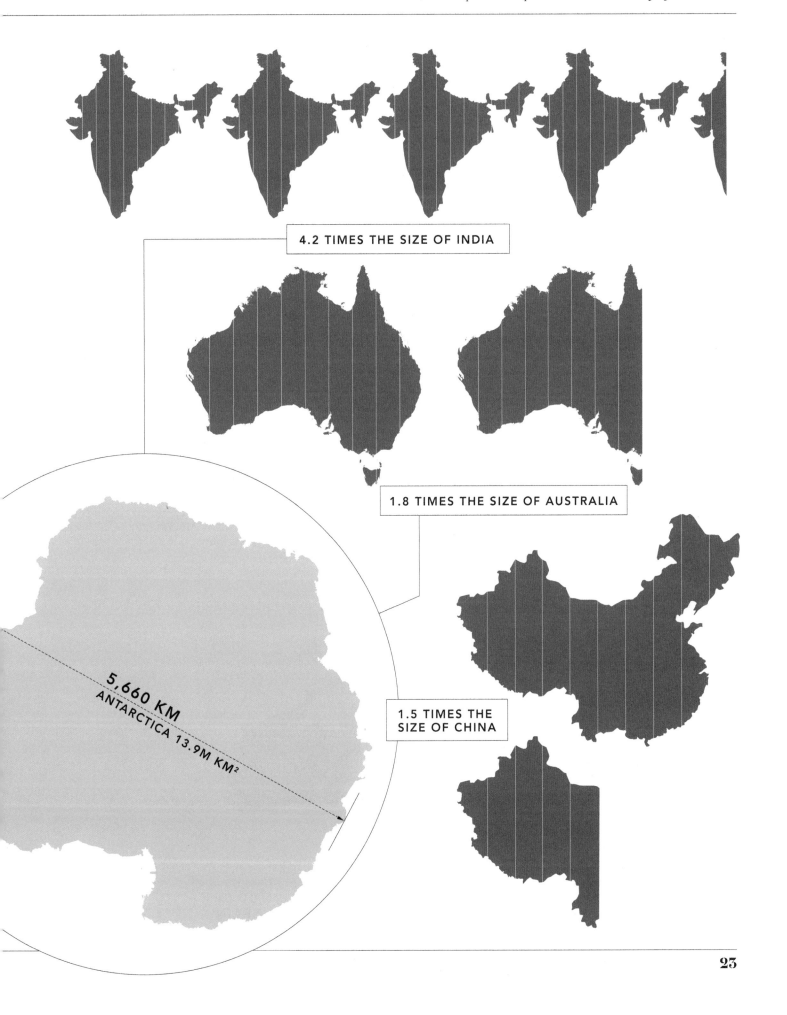

4.2 TIMES THE SIZE OF INDIA

1.8 TIMES THE SIZE OF AUSTRALIA

1.5 TIMES THE SIZE OF CHINA

5,660 KM
ANTARCTICA 13.9M KM²

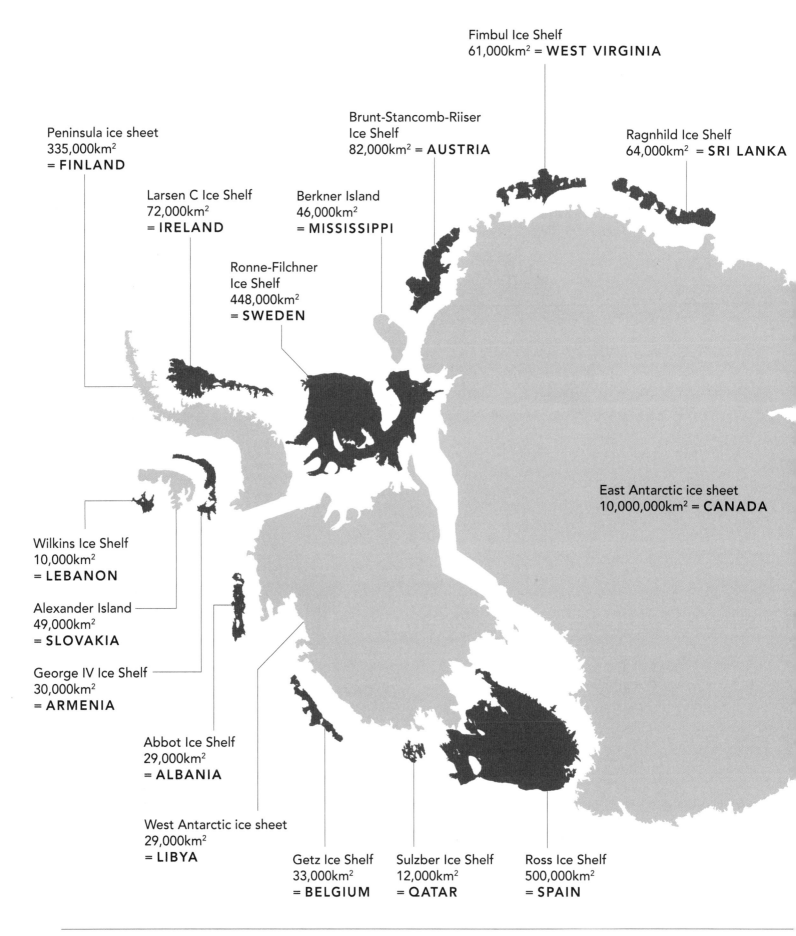

Fimbul Ice Shelf
61,000km² = **WEST VIRGINIA**

Brunt-Stancomb-Riiser
Ice Shelf
82,000km² = **AUSTRIA**

Ragnhild Ice Shelf
64,000km² = **SRI LANKA**

Peninsula ice sheet
335,000km²
= **FINLAND**

Larsen C Ice Shelf
72,000km²
= **IRELAND**

Berkner Island
46,000km²
= **MISSISSIPPI**

Ronne-Filchner
Ice Shelf
448,000km²
= **SWEDEN**

East Antarctic ice sheet
10,000,000km² = **CANADA**

Wilkins Ice Shelf
10,000km²
= **LEBANON**

Alexander Island
49,000km²
= **SLOVAKIA**

George IV Ice Shelf
30,000km²
= **ARMENIA**

Abbot Ice Shelf
29,000km²
= **ALBANIA**

West Antarctic ice sheet
29,000km²
= **LIBYA**

Getz Ice Shelf
33,000km²
= **BELGIUM**

Sulzber Ice Shelf
12,000km²
= **QATAR**

Ross Ice Shelf
500,000km²
= **SPAIN**

# 6. *The shelf and the sheet*

**THE MOST IMPORTANT THING** to understand about Antarctic geography is the difference between an ice shelf and an ice sheet. Often misquoted and misunderstood, both terms refer to types of glacial ice – ice that has formed on land from compacted snow – but other than that, there are a number of subtle differences between the two.

The ice sheet covers the centre of the continent and its base rests on the rock beneath it. It is usually very thick, averaging around 2 kilometres, and although some areas of ice are relatively fast-moving (usually termed 'ice streams'), on the whole its ice moves slowly.

By comparison, ice shelves are situated around the periphery of the land mass. They are fast-moving and much thinner – never over a kilometre thick – but the key distinction is that ice shelves are floating. Ice floats when you put it in water (think of the ice cubes in a drink), but the difference in density between ice and liquid water is small, so only around one tenth of an ice cube, or an ice shelf, sticks out of the water. When an ice sheet reaches the coast, if less than one tenth of the ice is above sea level, its base, which was grounded to the earth, detaches from its rocky bed and the whole ice mass, often hundreds of metres thick, starts to float. It is at this point that it becomes an ice shelf. The line where the ice separates from the ground is usually termed the 'grounding line'.

This map shows the different ice shelves and ice sheets around Antarctica. The ice sheets are usually divided into three: the East, the West and the Peninsula ice sheets. Each is massive; the East Antarctic ice sheet alone is as large as Canada, the world's second-largest country. The ice shelves are smaller, but each is still as large as a medium or small nation. The largest, the Ross Ice Shelf, is around the size of Spain.

Amery Ice Shelf
63,000km² = **LATVIA**

West Ice Shelf
21,000km² = **EL SALVADOR**

Shackleton Ice Shelf
33,000km² = **BELGIUM**

**ICE SHEETS**

**ICE SHELVES**

# 7. The day, the night and the half-light

**TIME IS TRULY** an abstract concept in Antarctica. If you define a day as the period between when the sun rises, sets and rises again, then at the South Pole one day lasts for a whole year. At the Pole, the sun rises into view in late September, after which there is continuous daylight for the next six months. Then the sun drops below the horizon, marking the start of six months of total darkness. The equinoxes are times of twilight, when the disc of the sun brushes the horizon in its 360-degree rotation around the Earth. As you travel northwards, away from the Pole, you start to get 24-hour periods when part of the day is in sunlight and part is in darkness – what we might call 'true days'. This happens first around the spring and autumn (vernal) equinox. In late March and September, at 89 degrees south, instead of tracking around the horizon, the sun will dip below and rise above it for a few days a year. Further north, away from the Pole, the number of true days increases and the 24-hour periods of total darkness and total light become fewer, until at 66°33' 6.6" south, there is never a day of total darkness or total light. This line is called the 'Antarctic Circle'.

On the facing page is a time map, a map that juxtaposes time with latitude, showing the times of the year when there are periods of total daylight (light yellow) and total darkness (blue). The circles on the map indicate the distance from the South Pole in divisions of two degrees of latitude (about 220 kilometres per circle). Around the circle are the months of the year, running clockwise from January to December. The latitudes of several Antarctic research stations are indictated in the upper left quadrant on the concentric circles radiating out from the Pole.

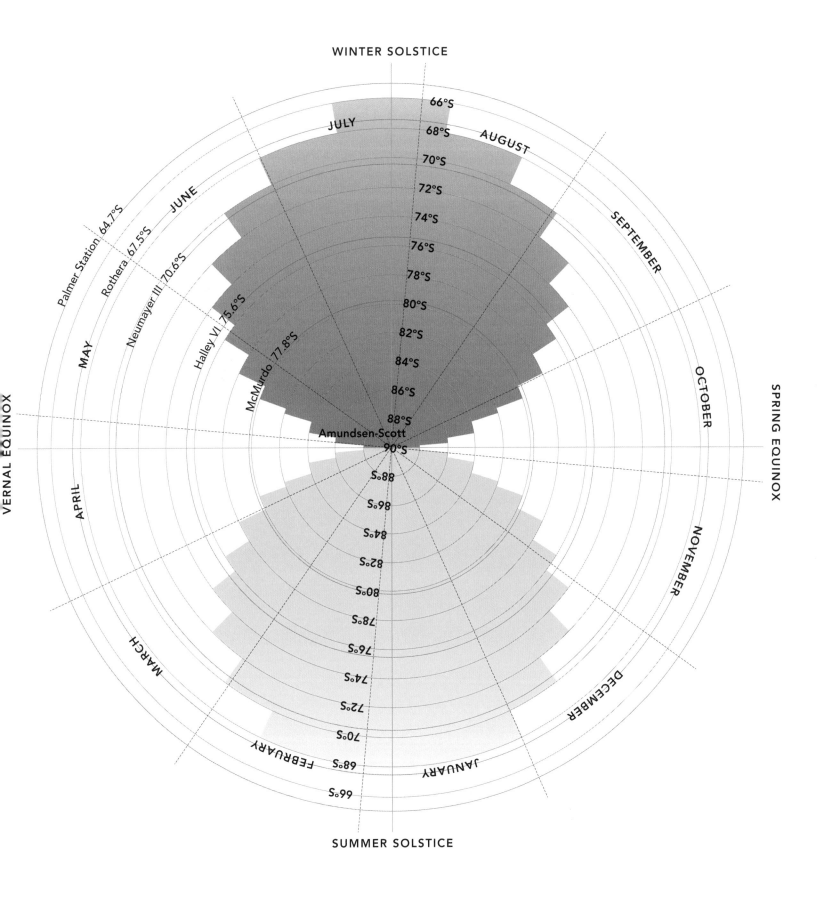

WINTER SOLSTICE

JULY AUGUST

JUNE SEPTEMBER

66°S
68°S
70°S
72°S
74°S
76°S
78°S
80°S
82°S
84°S
86°S
88°S

Palmer Station 64.7°S
Rothera 67.5°S
Neumayer III 70.6°S
Halley VI 75.6°S
McMurdo 77.8°S
Amundsen-Scott 90°S

MAY

APRIL OCTOBER

VERNAL EQUINOX SPRING EQUINOX

MARCH NOVEMBER

88°S
86°S
84°S
82°S
80°S
78°S
76°S
74°S
72°S
70°S
68°S
66°S

FEBRUARY DECEMBER

JANUARY

SUMMER SOLSTICE

# CHAPTER 2

# *Ice*

8. The ice beneath your feet *30*

9. Ice flow *32*

10. A changing world *34*

11. Drowning coasts *36*

12. The anatomy of an ice sheet *38*

13. Draining the depths *40*

14. Antarctic time machine *42*

15. Shrinking shelves *44*

# 8. *The ice beneath your feet*

**IN THIS CORNER** of the Earth, ice reigns supreme. Snow and ice are everywhere, and frozen water, in all its many guises, defines the continent. The ice sheet is on average around 2 kilometres thick, but this depth of ice is not evenly distributed; even near the heart of the continent, the ice can be quite thin, with the peaks of submerged mountains breaking through to the surface. In other regions, the ice sheet can be almost 5 kilometres thick. Huge, deep basins exist under the West and East Antarctic ice sheets, linked to some of the largest glacier catchment areas. These catchments (page 33) contain much of the volume of the ice sitting on the continent. Four of them – the Lambert, Byrd, Recovery and Totten catchments – contain 40 per cent of all the ice volume in Antarctica. The map on the facing page shows the distribution of ice thickness, highlighting a number of these deep basins and subglacial troughs, where the ice is thickest, in dark hues, with lighter blues indicating thinner ice.

But exactly how much ice is there? The calculation to estimate the amount is fairly simple: the ice sheet averages around 2 kilometres thick, and the continent is 14 million square kilometres, so that makes 28 million cubic kilometres of ice. Considering that a cubic metre of ice weighs almost a tonne, it equates to a grand total of 26 quadrillion tonnes of ice – that's 26 with fifteen zeros after it, enough mass pressing down on the Earth's crust under the ice to depress the underlying rock by almost a kilometre.

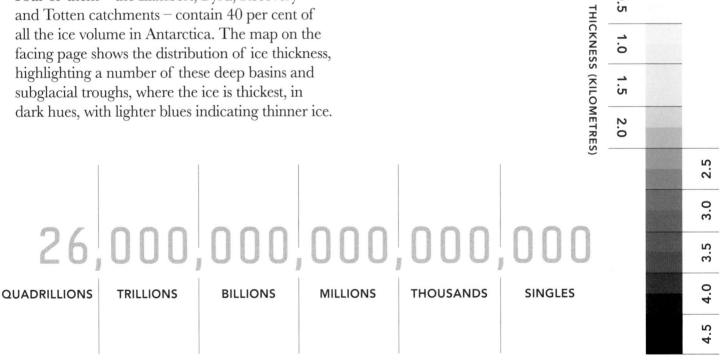

26,000,000,000,000,000

QUADRILLIONS | TRILLIONS | BILLIONS | MILLIONS | THOUSANDS | SINGLES

ICE THICKNESS (KILOMETRES)

0.5  1.0  1.5  2.0  2.5  3.0  3.5  4.0  4.5

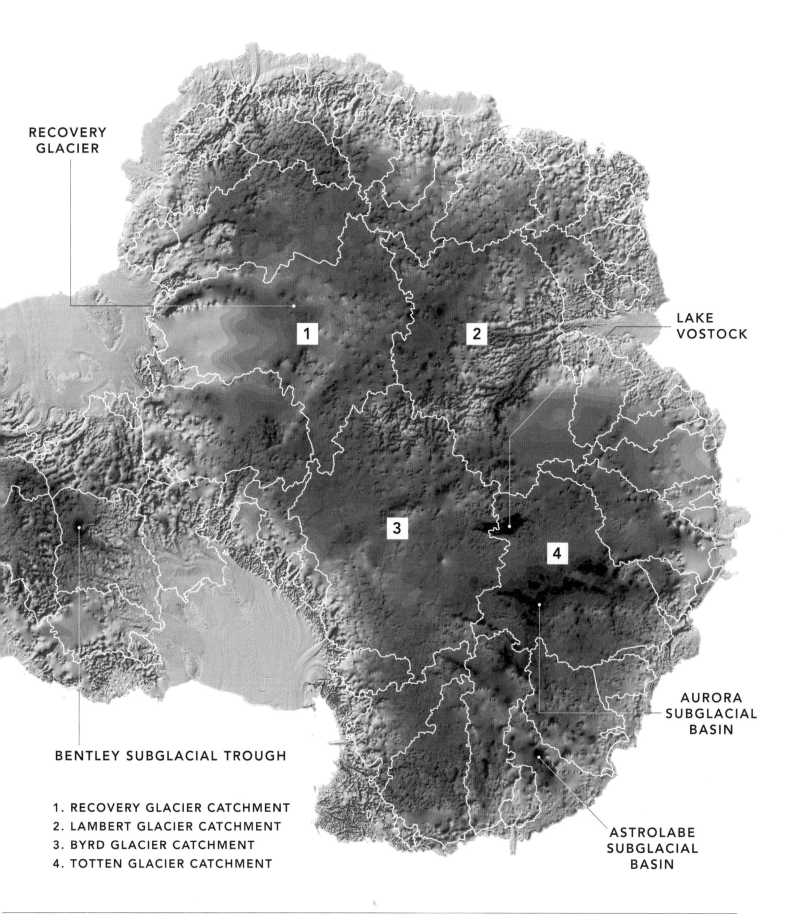

RECOVERY
GLACIER

LAKE
VOSTOCK

**1**

**2**

**3**

**4**

AURORA
SUBGLACIAL
BASIN

BENTLEY SUBGLACIAL TROUGH

1. RECOVERY GLACIER CATCHMENT
2. LAMBERT GLACIER CATCHMENT
3. BYRD GLACIER CATCHMENT
4. TOTTEN GLACIER CATCHMENT

ASTROLABE
SUBGLACIAL
BASIN

# 9. Ice flow

**ANTARCTICA IS ALIVE**. Almost the whole continent is continually moving, flowing like a frozen river downwards towards the coast. Imperceptibly slow, moving barely millimetres per year in the centre of the continent, the ice gradually gathers speed, accelerating as it moves downhill, deforming and compressing as it progresses, merging into huge ice streams that drain the land mass. By the time the ice reaches the sea it can be moving at up to 4 kilometres per year.

Each glacier drains a particular area – its catchment. The largest of these catchments are as big as South Africa or Colombia. Each of the four largest – the Lambert, Byrd, Recovery and Totten – covers an area of almost a million square kilometres, and together they drain ice from a third of the continent.

The extreme cold in Antarctica means that the ice at the surface never melts. The only way for the ice sheet to lose ice is to shed it in the form of icebergs at the coast. The continual snowfall and lack of melt result in more and more snow piling up, thicker and thicker, until the weight on the snowpack at the top compresses the light, fluffy flakes, turning them into solid ice. As more snow is added and the thickness increases further, the massive pressure from above starts to squeeze the ice crystals together, making them slide over each other. At that point, the ice starts to flow. The speed of that flow is determined mainly by the steepness of the slope – the steeper the gradient, the faster the flow. In the centre of the land mass the surface is virtually flat, but near the coast it is much steeper, giving Antarctica the profile of an upturned bowl. In the places where catchments meet (the white dashed lines on the map), the speed of flow is virtually zero. These lines are termed 'ice divides'.

Normally, the ice is in balance, that is, there is as much ice calving off as glaciers around the coast as there is new ice being made by snowfall in the interior. However, as the climate changes, the amount of snowfall varies. Too much snow will result in higher accumulation and, as the ice does not melt, eventually a higher ice surface. This will lead to a steeper slope, which in turn will increase the speed of the glaciers, which means more icebergs. So, normally, the ice regains its balance and equality is maintained. The problem comes when warm seawater melts the ice at the coast, causing ice fronts that retreat inland. When this happens, the gradient of the slope increases, leading to faster flow and more calving off that is not replaced by snowfall, and a run-away process of ice melt begins.

Scientists fear that in some areas of West Antarctica, such as Pine Island and Thwaites glaciers, this process has already started. As you can see on the map, these areas have the fastest-flowing glaciers. Pine Island Glacier is already losing more ice than any other glacier on the planet. It alone is responsible for 25 per cent of Antarctica's total ice loss and is raising global sea levels by one fifth of a millimetre a year. The extra water added to the world's oceans from these, and other, similar ice streams could raise sea levels by over a metre by the end of this century, putting many cities, coastal communities and small islands at risk.

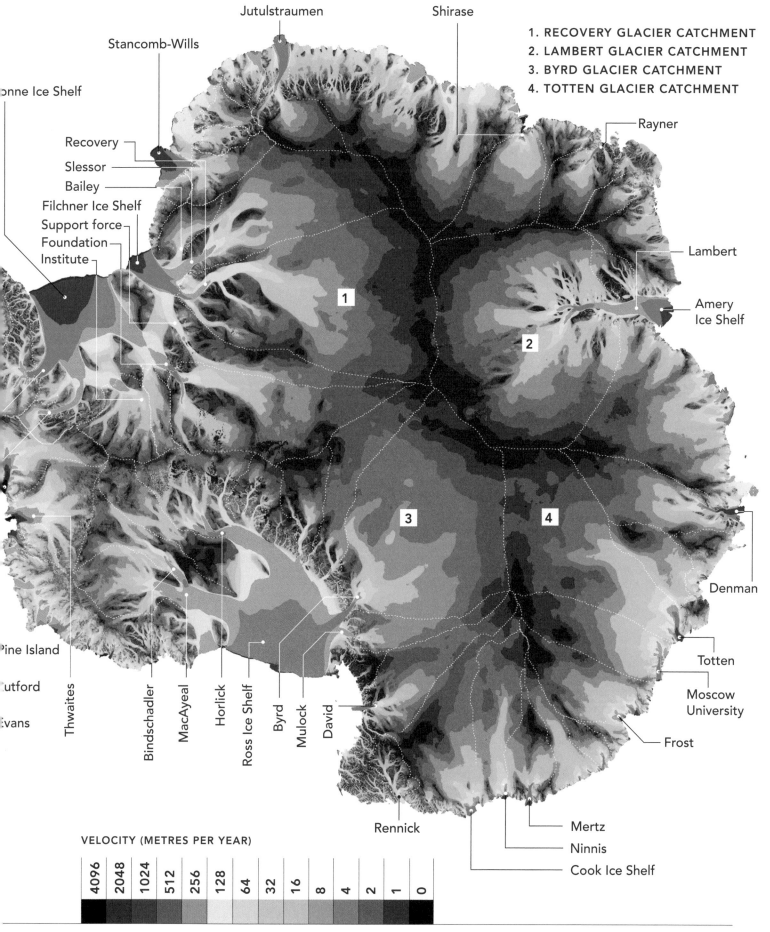

1. RECOVERY GLACIER CATCHMENT
2. LAMBERT GLACIER CATCHMENT
3. BYRD GLACIER CATCHMENT
4. TOTTEN GLACIER CATCHMENT

Jutulstraumen

Shirase

Stancomb-Wills

onne Ice Shelf

Recovery

Slessor

Bailey

Filchner Ice Shelf

Support force

Foundation

Institute

Rayner

Lambert

Amery
Ice Shelf

Pine Island

utford

vans

Thwaites

Bindschadler

MacAyeal

Horlick

Ross Ice Shelf

Byrd

Mulock

David

Rennick

Denman

Totten

Moscow
University

Frost

Mertz

Ninnis

Cook Ice Shelf

**VELOCITY (METRES PER YEAR)**

| 4096 | 2048 | 1024 | 512 | 256 | 128 | 64 | 32 | 16 | 8 | 4 | 2 | 1 | 0 |
|------|------|------|-----|-----|-----|----|----|----|---|---|---|---|---|

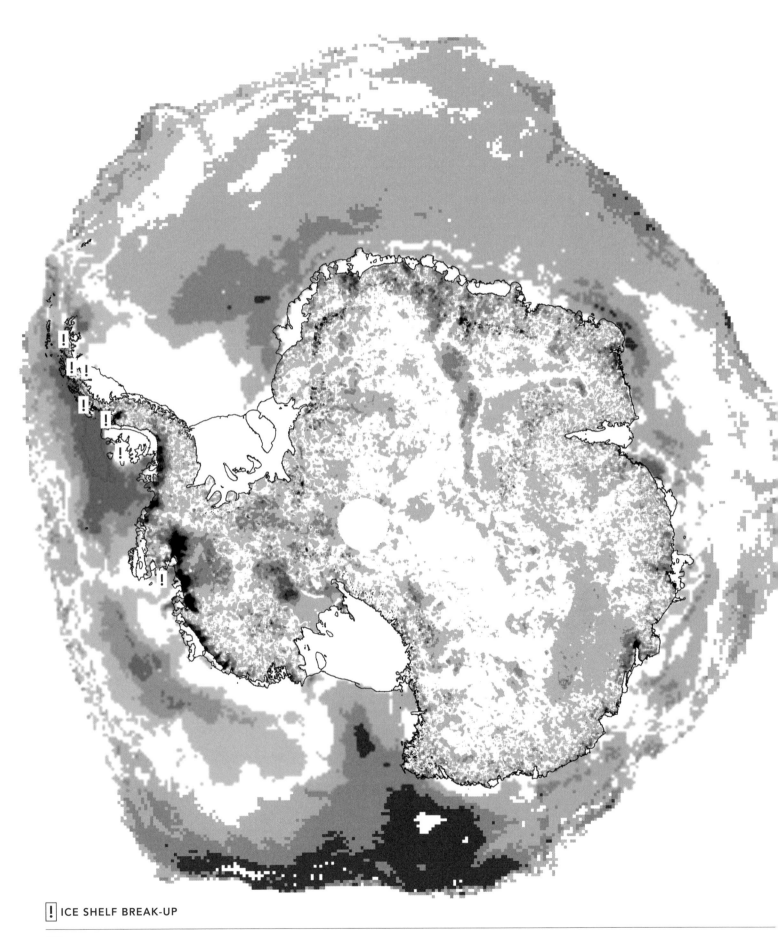

!  ICE SHELF BREAK-UP

# 10. A changing world

**ANTARCTICA IS CHANGING**. As the world warms, patterns of change manifest themselves on this frozen continent, but, unlike in the Arctic, where ice is melting everywhere, the signal in the Antarctic is not so clear.

Yes, there is warming: in parts of the Antarctic Peninsula, temperatures are rising faster than virtually anywhere else on the globe. Glaciers in the Amundsen Sea in West Antarctica are melting swiftly and contributing more water to global sea-level change than any other glaciers on the planet. Sea ice around the Peninsula is disappearing, and ice shelves in the area are collapsing like dominoes. However, in East Antarctica, the pattern is reversed. Here, the ice sheets are gaining mass; in some areas, the temperatures are cooling and the sea ice is growing (see page 83 for more details on this interesting phenomenon). Overall, the pattern is unclear; there are significant losses in some parts of the continent and observable gains in others. With one cancelling out the other, the balance is almost neutral, with a slight overall loss in the amount of glacial ice and a slight increase (about 1 per cent a decade) in the extent of sea ice.

The map shows the pattern of changes. The red areas show loss – loss of sea ice offshore and glacial ice onshore – while the blue areas show increasing amounts of ice. For sea ice, the amount of change in ice coverage per year is mapped. In some areas, four days' worth of sea-ice cover is lost each year. For land ice, on the ice sheets and glaciers, the change in altitude of the ice sheet is shown. Here, too, loss – up to 2 metres a year – outweighs gain. The areas with the greatest loss of sea ice, around the Amundsen Sea, near the mouths of Pine Island and Thwaites glaciers, are also the areas with largest loss of land ice. Similarly, the areas gaining most sea ice, in the Ross Sea sector, are also the areas that are gaining ice on land.

These changes have consequences for us all. Ice melting from the land adds to global sea level (see page 36), while changes in sea-ice cover can disturb global ocean circulation (see page 98) and have devastating consequences on local wildlife (see page 110).

**TREND IN THE ICE SEASON DURATION
1979/80 TO 2011/12 (DAYS PER YEAR)**

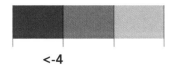 

<-4          0          >4

**ICE SHEET ELEVATION CHANGE (METRES PER YEAR)**

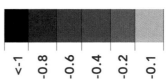 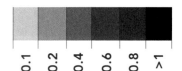

<-1  -0.8  -0.6  -0.4  -0.2  -0.1          0.1  0.2  0.4  0.6  0.8  >1

# 11. Drowning coasts

**IF ANTARCTICA MELTS**, the sea level rises, and this is not an effect that will be confined locally. Ice melting off Antarctica raises the level of the oceans in the northern hemisphere as much as it raises the levels of those in the south. If all the ice in Antarctica melted, the sea level around the world would rise by 58 metres, flooding eastern England, the Netherlands, Bangladesh and many other low-lying coastal areas around the globe. Add on the ice melt from Greenland and mountain glaciers, and the potential sea-level rise from glacial ice is over 66 metres. So all areas that lie below 66 metres above sea level would, in theory, be inundated. The areas shown on the map in dark red lie below this limit. Luckily, much of Antarctica is unlikely to melt any time soon; even in the worst future climate-change scenarios, the vast ice caps of East Antarctica will take thousands of years to melt.

The more vulnerable parts of the continent – West Antarctica and the Antarctic Peninsula – are more susceptible to rapid change. They, along with Greenland and mountain glaciers, are already undergoing a transformation that many scientists think is irreversible. Under current projections of global temperature change, much of the ice from these regions could melt within a few centuries. That would add around 13 metres to global sea-level rise; the areas shown on the map in lighter red are below this level and are vulnerable to future flooding. Many coastal areas will be affected by this change. Some low-lying countries could be inundated, and many regions will be threatened, including eastern England, Florida, tropical islands

such as the Maldives, Bangladesh and parts of China, all of which have large vulnerable low-lying areas. Also, in China, Florida and Bangladesh, very large populations live in these regions.

And the map does not show the full extent of sea-level rise. It is worse than that. This map shows an equal rise across the oceans, but areas nearer the equator will see a higher rise than those closer to the poles. The billions of tonnes of ice in the Antarctic ice cap have a gravitational attraction, and as they melt, the mass of the water will redistribute into the oceans, the attraction to the poles will reduce and water will spread unevenly towards the tropics. Consequently, flooding in Bangladesh, China, Florida and some tropical islands will be worse than is indicated here.

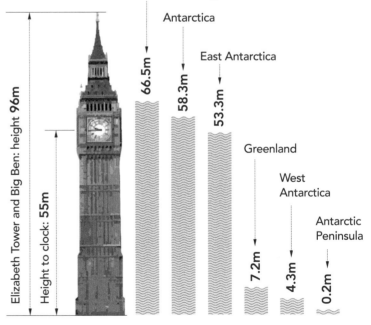

Total potential sea-level rise from Antarctica, Greenland and mountain glaciers

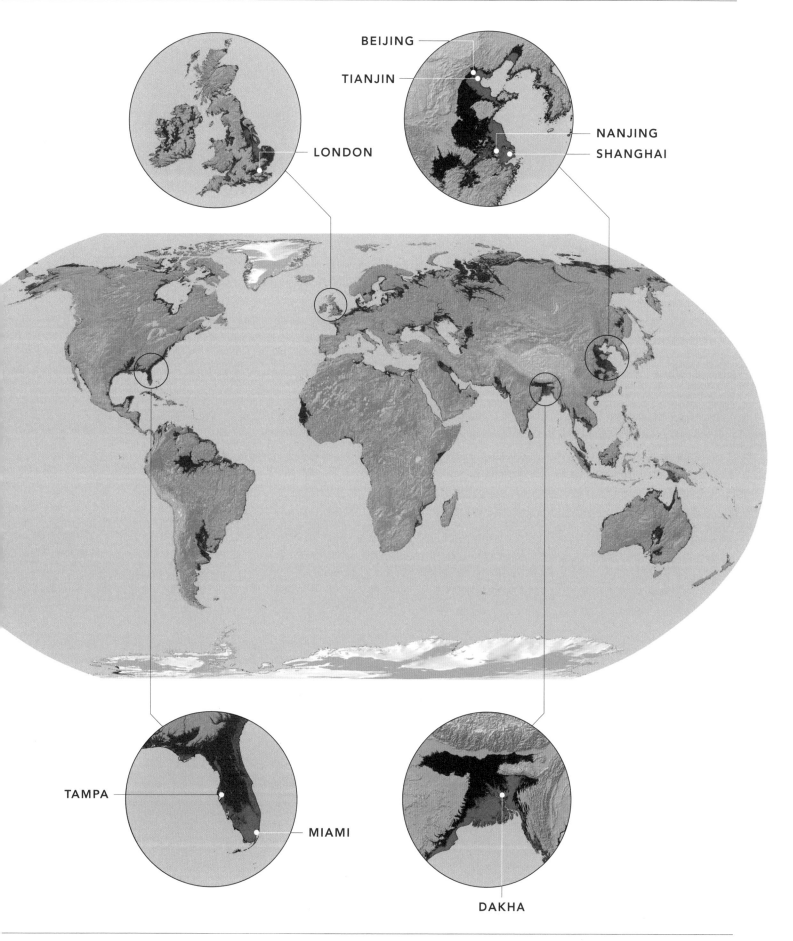

BEIJING

TIANJIN

LONDON

NANJING

SHANGHAI

TAMPA

MIAMI

DAKHA

# 12. The anatomy of an ice sheet

**ALMOST THE WHOLE** surface of Antarctica is encased in a permanent frozen coat of ice and snow, with only a tiny portion (around a quarter of 1 per cent) exposed as rock. The ice takes on many forms and provides a blank canvas on which the weak Antarctic sun paints a myriad of hues and colours, from the cobalt blues of the deep ice to the pastel shades of pink and ochre in which the sun daubs the sea ice as it skims the horizon. The ice can be beautiful, but it can also be deadly. In places, it is riven by crevasses, massive cracks that penetrate hundreds of metres into its heart. Travelling over such treacherous terrain is a risky business and has been the end of many an Antarctic explorer – or scientist, for that matter.

The ice is dynamic and ever-changing. As snow falls, it settles and is covered by more snow. This snow is gradually compressed and converted into ice. Over time, the pressure from more and more snow and ice above deforms the ice crystals and the ice starts to move. Gradually, often imperceptibly slowly, the ice moves downhill, flowing for hundreds, perhaps thousands, of kilometres before it reaches the coast. A single snowflake falling in the centre of the continent may take a million years or more before it rejoins the sea.

Sometimes only the top of a mountain sticks out from the ice. These features are called 'nunataks'.

When the ice reaches the sea, it spreads out over the ocean as a floating ice shelf, nine-tenths of which is below sea level.

When the ice shelf becomes unstable, it calves off into icebergs, each often many kilometres wide.

Beyond the continental margin, the continental slope drops off steeply to the abyssal plain.

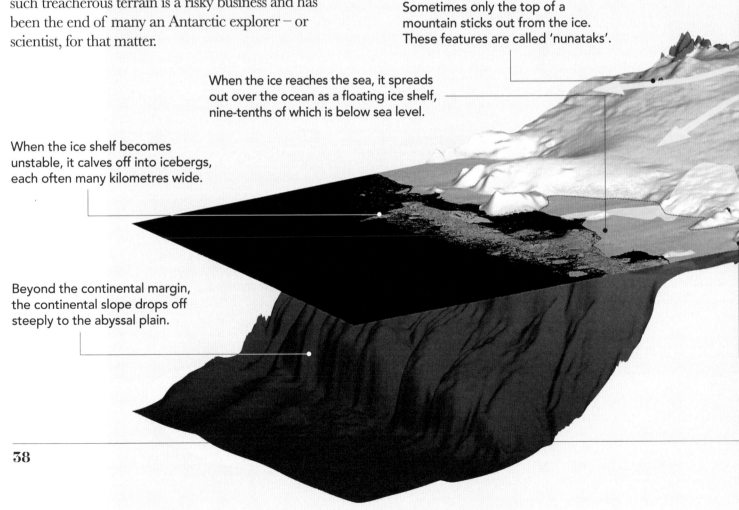

Often, at the edge of the continent, mountain ranges hold back the ice sheets, damming the great volumes of ice. Here, the ice streams must cut through the mountains, punching wide, deep channels through the rock. Eventually, after thousands of kilometres and tens of thousands of years, the ice reaches the sea, but even here its journey does not end. The temperature of the seawater around the coast is below zero degrees, so the ice does not melt, and ice streams flow out into the sea. As ice is less heavy than water, buoyancy causes the ice to float, and its base detaches from the rocky bed. These floating ice tongues merge with other glaciers to form massive floating ice shelves. Freed from the drag of the land, the ice flows faster, often moving at speeds of several metres a day, until, finally, buffeted by storms and tides, the end of the shelf cracks and the extremities of the ice break away to form huge, tabular bergs.

When the snow falls it is buried beneath the ice sheet, but the ice sheet is moving so the snow's path is both downwards and towards the coast.

The ice is constantly moving downhill. This movement is extremely slow in the interior, but accelerates as the ice gets closer to the coast.

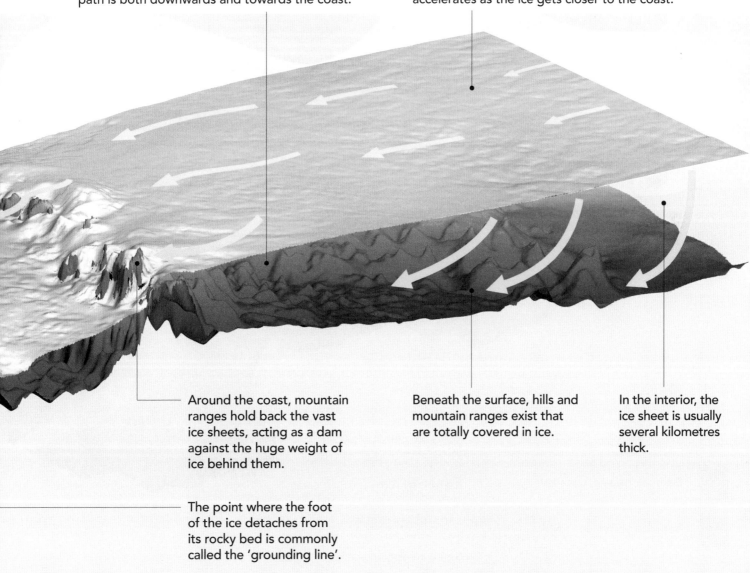

Around the coast, mountain ranges hold back the vast ice sheets, acting as a dam against the huge weight of ice behind them.

Beneath the surface, hills and mountain ranges exist that are totally covered in ice.

In the interior, the ice sheet is usually several kilometres thick.

The point where the foot of the ice detaches from its rocky bed is commonly called the 'grounding line'.

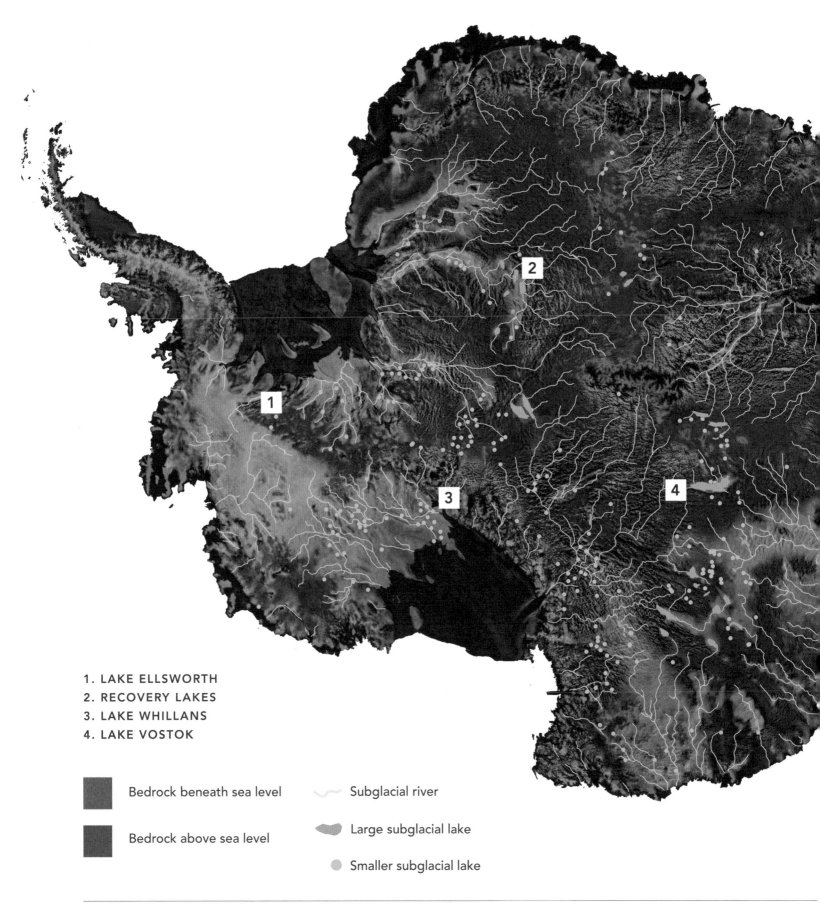

1. LAKE ELLSWORTH
2. RECOVERY LAKES
3. LAKE WHILLANS
4. LAKE VOSTOK

Bedrock beneath sea level

Bedrock above sea level

Subglacial river

Large subglacial lake

Smaller subglacial lake

# 13. Draining the depths

**DEEP BELOW THE ANTARCTIC** ice sheets there is often liquid water. A whole network of lakes and streams drains meltwater from the interior of the continent, running thousands of kilometres to the coast. It seems strange that in a place where the temperature at the surface often averages 50 degrees below zero, there can still be liquid water below your feet.

Why does this happen? Firstly, the great thickness of the ice acts as a superb insulator, never allowing the cold to be transferred to the bottom of the ice sheet, so the water below does not freeze. Secondly, the rock beneath the ice often radiates small amounts of geothermal heat. And thirdly, the weight of thousands of tonnes of ice above creates huge pressure. Water molecules find it harder to bond together into solid form when under pressure, so the weight of the ice and small amounts of heat mean that water can exist in liquid form well below zero degrees Celsius.

So what we have is a ghostly parallel of what might exist on a continent without ice sheets, with water, rivers and lakes beneath the ice. Some of these lakes are huge. The largest, Lake Vostok, is similar in size to Lake Ontario, and would be the fifteenth-largest lake in the world. Many other large lakes have been discovered, and many are connected by subglacial streams and rivers. Scientists can see changes in the flow of the lakes by subtle height differences on the surface of the ice. The map shows our current understanding of this drainage network beneath the ice.

With a surface area of 12,500km², Subglacial Lake Vostok is the world's largest known subglacial lake. This graphic shows a three-dimensional cross section through the lake and the ice above it.

- Ice surface: 3,520m above sea level
- Ice depth: 4,030m
- Lake surface: 510m below sea level
- Lake depth: 870m
- Bottom of lake: 1,380m below sea level

# *14. Antarctic time machine*

**IN ANTARCTICA**, you can go back in time. The ice contains a secret history of the Earth's climate that goes back hundreds of thousands of years. When snow falls in the Antarctic, it never melts, it just accumulates, snow on snow, year on year, until it is kilometres thick. As the weight of the ice crushes the fluffy snowflakes, it turns the snowpack first to a granular ice called 'firn', then, eventually, into solid glacial ice. The air that was originally trapped in the snow when it landed as a snowflake is entombed as small bubbles within the solid ice, and these bubbles contain a record of the atmosphere and climate at the time the snow originally fell.

As more snow falls, the older ice is buried deeper and deeper; so, generally, the deeper the ice, the older it is. But it's not quite that simple: the pressure of the tonnes of ice above deforms the ice crystals and the ice starts to flow, and the further from the centre of the continent it is, the faster it flows. This keeps the amount of ice in balance and also means that, towards the coasts, where the ice is thinner and faster-moving, it is younger. The oldest ice is always deep within the continental interior at the drainage divides, where the speed of the ice flow is almost zero.

**Mt. HADDINGTON
650M / 14,000 YEARS**

**SIPLE
200M / 250 YEARS**

The map shows the locations of the major ice cores taken on the continent, their depth and the age of the oldest ice at the bottom of the core. Over the past fifty years, these ice cores, along with similar (but shorter) cores from Greenland, have given us a detailed picture of the climate of the Earth over the past three quarters of a million years. Those tiny air bubbles deep within the ice have shown us how humanity has changed the global atmosphere and how carbon dioxide and other greenhouse gases have increased as we have burned more fossil fuels. The ice cores are the smoking gun that proves the greenhouse effect and the theory of human-induced climate change.

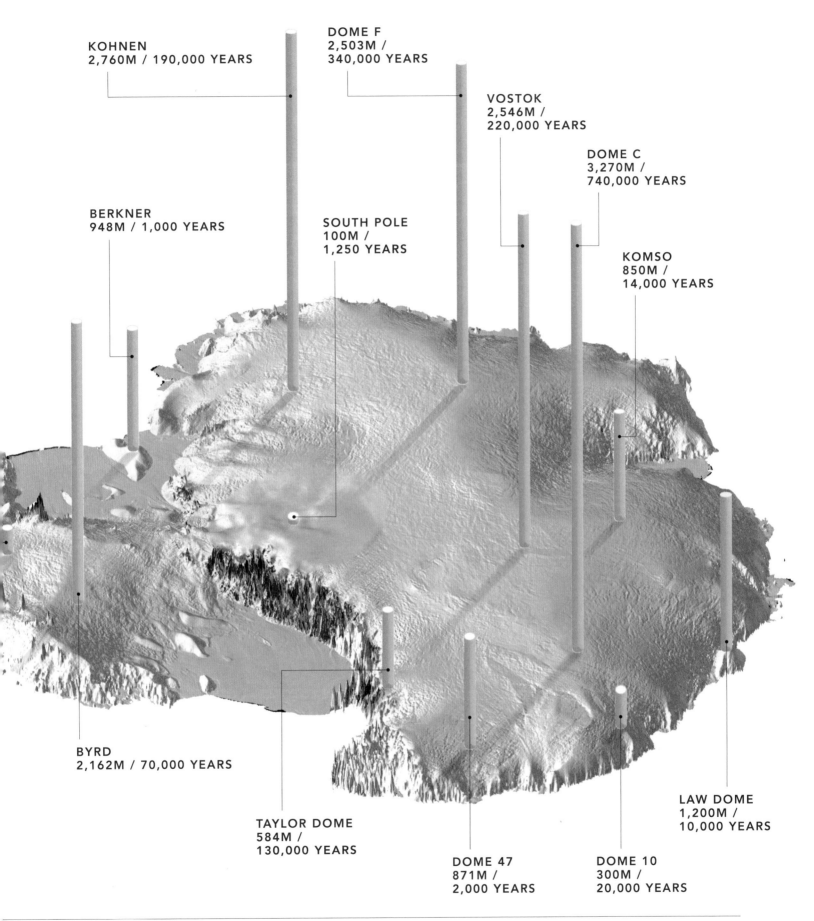

KOHNEN
2,760M / 190,000 YEARS

DOME F
2,503M /
340,000 YEARS

VOSTOK
2,546M /
220,000 YEARS

DOME C
3,270M /
740,000 YEARS

BERKNER
948M / 1,000 YEARS

SOUTH POLE
100M /
1,250 YEARS

KOMSO
850M /
14,000 YEARS

BYRD
2,162M / 70,000 YEARS

TAYLOR DOME
584M /
130,000 YEARS

DOME 47
871M /
2,000 YEARS

DOME 10
300M /
20,000 YEARS

LAW DOME
1,200M /
10,000 YEARS

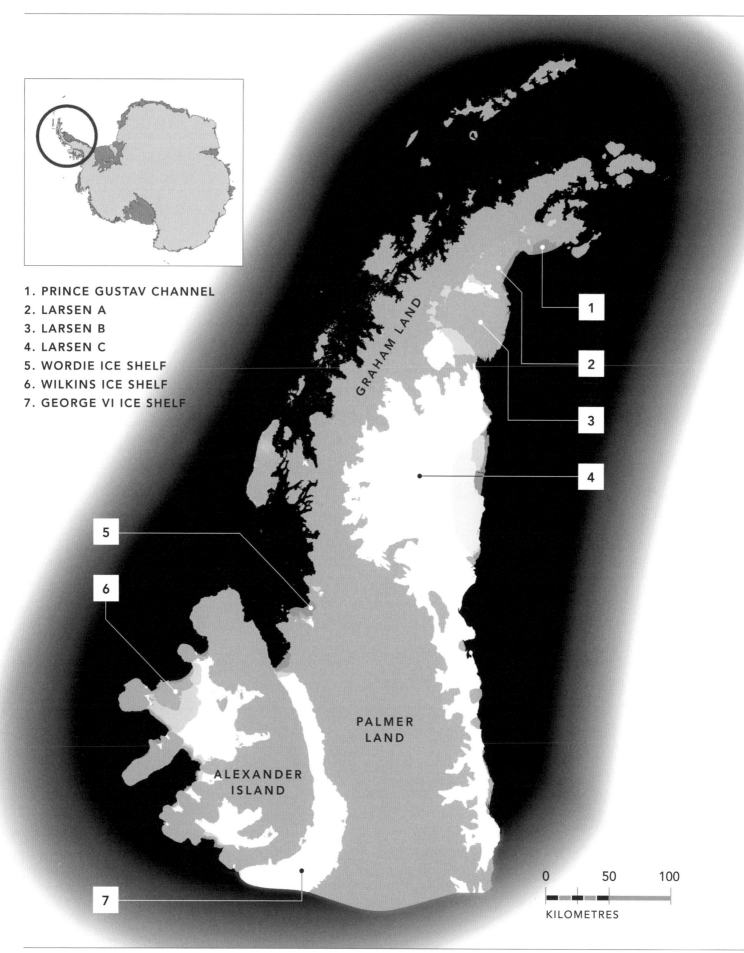

1. PRINCE GUSTAV CHANNEL
2. LARSEN A
3. LARSEN B
4. LARSEN C
5. WORDIE ICE SHELF
6. WILKINS ICE SHELF
7. GEORGE VI ICE SHELF

GRAHAM LAND

PALMER LAND

ALEXANDER ISLAND

0     50     100

KILOMETRES

# 15. Shrinking shelves

**ONE OF THE MOST** visible signs of climate change in Antarctica has been the disintegration of ice shelves on the Antarctic Peninsula. This map shows those changes, with darker blue areas indicating earlier ice losses and lighter blue areas showing more recent break-ups. The white areas indicate the shelves that still remain and the grey areas indicate land ice.

While most of the continent has not yet fallen victim to the rising temperatures that have affected many other parts of the globe, the Antarctic Peninsula, the least cold part of Antarctica, has seen dramatic warming. In some parts of the Peninsula temperatures have risen by over 4°C in the last fifty years, the greatest regional warming of any place on Earth. This change has had serious consequences. Glaciers have retreated, the amount of exposed rock has risen and ice shelves have collapsed. It started in the 1950s, with the retreat of the most northerly ice shelf, Prince Gustav Channel, which finally disintegrated in the 1980s. Since then, the trend has cascaded southwards, with the collapse of parts of the giant Larsen Ice Shelf, the third largest in Antarctica. First, the northern part, Larsen A, disintegrated, then Larsen B, a total of 3,250 square kilometres of ice. Now, only the largest part, Larsen C, remains, and this is showing increasing signs of instability, with a large chunk, measuring 5,800 square kilometres, breaking off in 2017.

Although the melting of ice shelves does not add to sea-level rise, losing them means that the glaciers and ice streams behind are no longer restrained. When the shelves disappear, the glaciers speed up and discharge more ice and water into the oceans, which in turn raises the sea level. Unlike ice shelves, the glaciers do add their water to the global sea level (see page 36). For this reason, the shelves have been termed the 'cork in the bottle' of Antarctic ice loss.

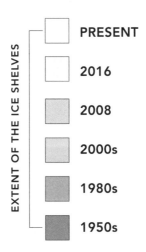

EXTENT OF THE ICE SHELVES

PRESENT

2016

2008

2000s

1980s

1950s

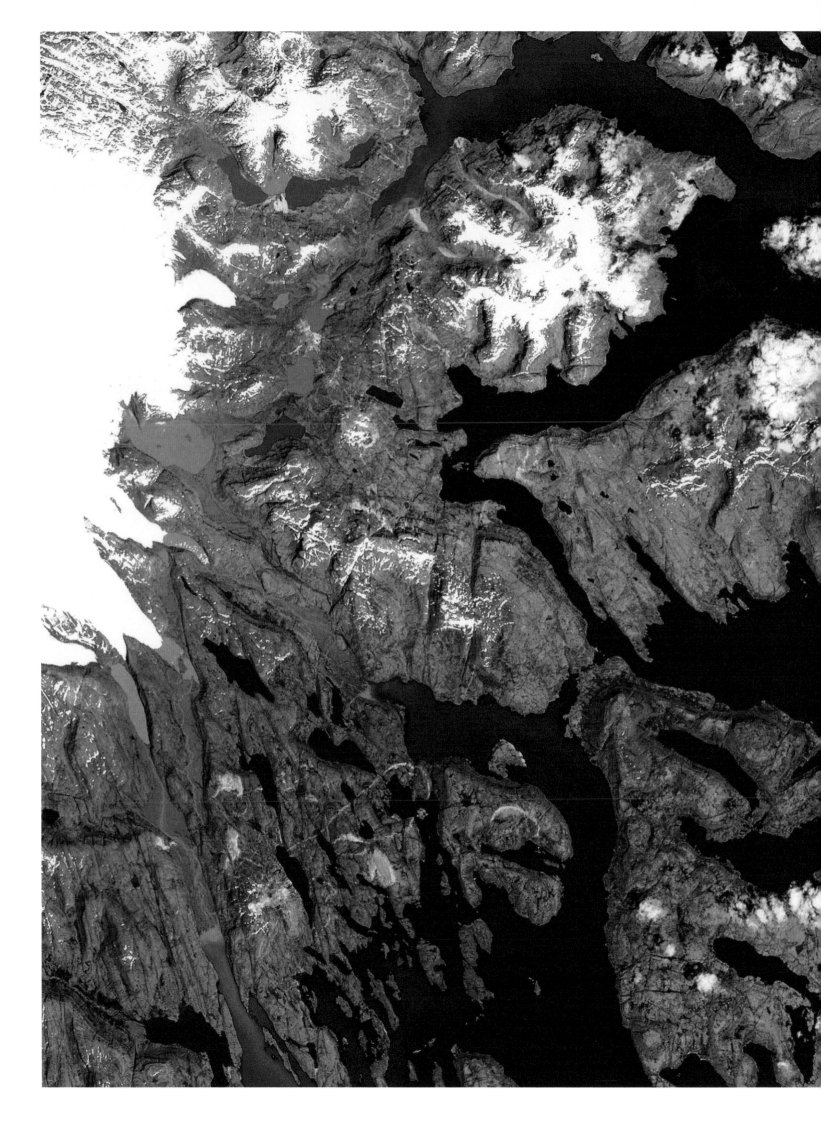

CHAPTER 3

# *Land*

16. Rocks below the ice  *48*

17. The hidden world  *50*

18. The making of Antarctica  *52*

19. Volcanoes  *54*

20. The quaking sea  *58*

21. The driest place on Earth  *60*

22. Alien invasion  *62*

23. Majestic mountains  *64*

24. The wolf at the end of the world  *68*

# 16. Rocks below the ice

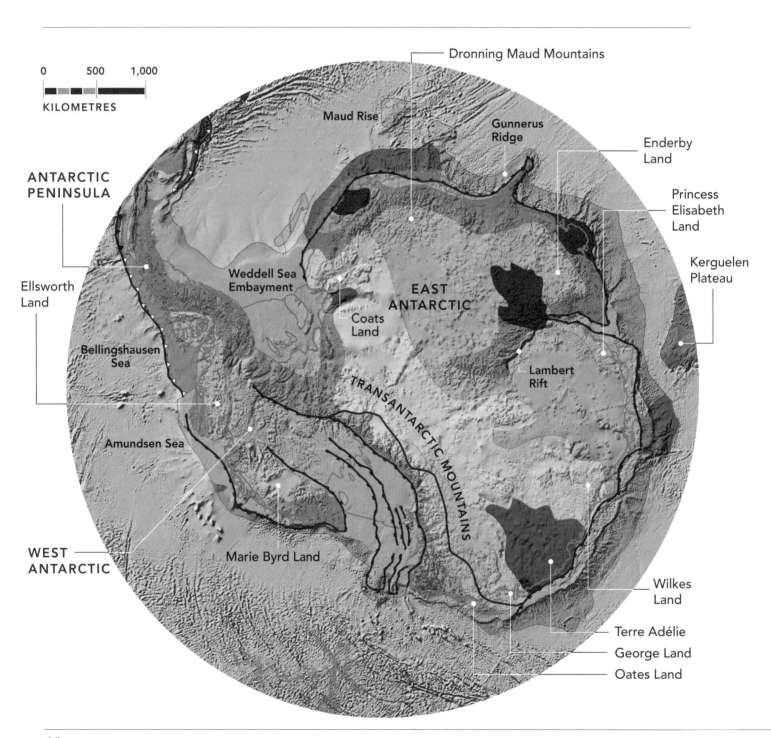

Dronning Maud Mountains

Maud Rise

Gunnerus Ridge

Enderby Land

Princess Elisabeth Land

Kerguelen Plateau

ANTARCTIC PENINSULA

Ellsworth Land

Weddell Sea Embayment

EAST ANTARCTIC

Coats Land

Bellingshausen Sea

Lambert Rift

Amundsen Sea

TRANSANTARCTIC MOUNTAINS

WEST ANTARCTIC

Marie Byrd Land

Wilkes Land

Terre Adélie

George Land

Oates Land

0    500    1,000

KILOMETRES

**TRYING TO WORK** out the geology of Antarctica is a tricky business. Only a tiny percentage (around 0.25 per cent) of the continent's rocks stick out above the ice, and in the interior there are vast areas where there are no rocks at all, just kilometre after kilometre of ice, thousands of metres deep. The outcrops that do exist are often remote, inaccessible and surrounded by crevasses, making it difficult for geologists to study them. Trying to understand the rocks beneath the ice is even harder; scientists have to interpret the slight differences in gravity and magnetics in their attempts to understand these rocks, which are totally unreachable. One thing that does help is the absence of soil and vegetation. In more temperate climates, bare rock is usually covered by thick layers of soil and hidden by trees and shrubs, but

here, in the coldest place on Earth, there is virtually no soil, and there are no plants except for lichens and the occasional patch of moss, so the rocks that are exposed are bare and fresh.

This map shows the best reconstruction of the geology above and below the ice and represents our current understanding of the geological setting of the continent. The continent can generally be divided into two parts: the old East Antarctica, and the young West Antarctic and Antarctic Peninsula. We think that East Antarctica is ancient – hundreds of millions of years old. However, most of this part of the continent is below the deepest ice, so much of the geology remains unknown. West Antarctic is younger, mostly created in the last 200 million years, and has had a very different geological history.

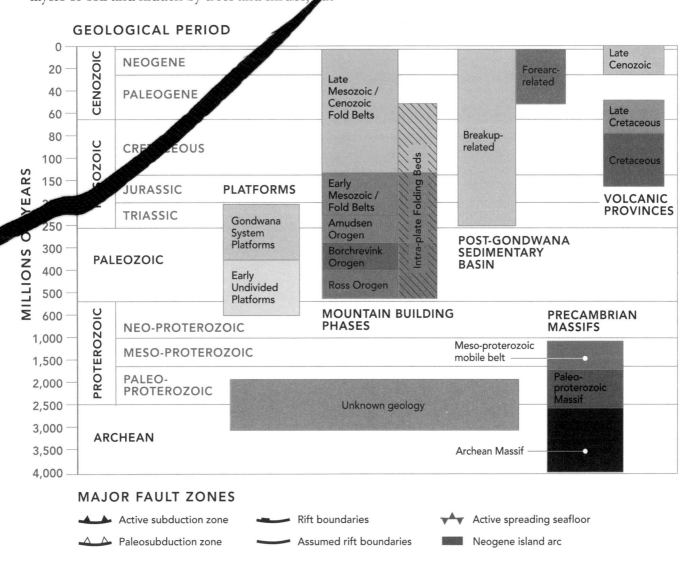

# 17. The hidden world

**HIDDEN BENEATH THE ANTARCTIC** ice sheet lies a secret world. The surface of the icy continent may appear flat and smooth, but under it lies a complex terrain, a landscape of mountains, rivers, lakes, hills, valleys and plains much like on any other continent. Yet no one will ever see this land. It is buried, entombed under up to 4 kilometres of ice which hides mountain ranges the size of the Alps, while in other areas gorges longer than the Grand Canyon drain water from the interior. Occasionally, where the ice is thinner, mountain peaks, known as nunataks, poke out through the surface.

This map shows the rocky surface underneath the ice. The colours relate to the height: blue is rock below sea level, yellow is above sea level, and the red areas indicate the highest mountains.

BedMachine Antarctica, the most recent scientific publication informing us of the terrain beneath the ice, is a compilation of many different survey techniques – radio-echo sounding, seismic data, gravity modelling and offshore sonar – that have been brought together to form a single dataset, on which this map is based. This information is vital for ice-sheet models, which in turn help us estimate how much of Antarctica might melt in years to come.

## MAP LOCATIONS

1. Shackleton Range
2. Pensacola Mountains
3. Ellsworth Mountains
4. Ellsworth Subglacial Highlands
5. Bransfield Strait
6. Peter I Øy
7. De Gerlache Seamounts
8. Pine Island Glacier
9. Thwaites Glacier
10. Executive Committee Range
11. Flood Range
12. Maudhelmvidda

13. Thiel Trough
14. Recovery Lakes
15. Gamburtsev Subglacial Mountains
16. Vostok Subglacial Highlands
17. Ross Island
18. Mawson Bank
19. Iselin Bank
20. Scott Seamounts
21. Balleny Islands
22. Gunnerus Ridge
23. Prince Charles Mountains
24. Amery Basin

25. West Lambert Rift
26. East Lambert Rift
27. Kerguelen Plateau
28. Subglacial Lake Vostok
29. Denman Glacier
30. Aurora Subglacial Basin
31. Totten Glacier
32. Resolution Subglacial Highlands
33. Astrolabe Subglacial Basin
34. Wilkes Subglacial Basin
35. Byrd Glacier

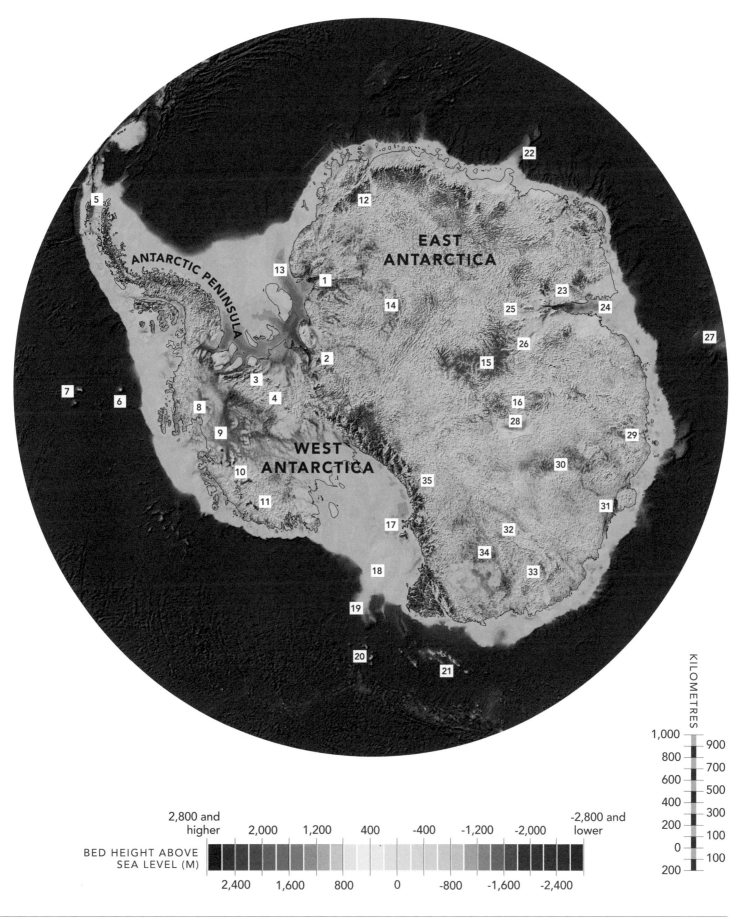

EAST
ANTARCTICA

ANTARCTIC PENINSULA

WEST
ANTARCTICA

| 1 | 2 | 3 | 4 | 5 | 6 | 7 | 8 | 9 | 10 | 11 | 12 | 13 | 14 | 15 | 16 | 17 | 18 | 19 | 20 | 21 | 22 | 23 | 24 | 25 | 26 | 27 | 28 | 29 | 30 | 31 | 32 | 33 | 34 | 35 |

KILOMETRES

1,000
900
800
700
600
500
400
300
200
100
0
100
200

| 2,800 and higher | | 2,000 | | 1,200 | | 400 | | -400 | | -1,200 | | -2,000 | | -2,800 and lower |

BED HEIGHT ABOVE
SEA LEVEL (M)

2,400      1,600      800      0      -800      -1,600      -2,400

# 18. The making of Antarctica

**THE ROCK OF ANTARCTICA** was not always covered in ice, nor was it always located at the bottom of the world. Once, it was part of a great supercontinent called Gondwana that included all the southern lands, including South America, Africa, India and Australia. Through the slow, inexorable pressure of plate tectonics, around 160 million years ago the immense land mass started to break apart. The Peninsula was joined to the Andes Mountains, part of a huge mountain chain – its rocks are still very similar to the mountains of South America today. Initially, East Antarctica was joined to Southern Africa, India, Australia and New Zealand, but these continental plates eventually split off and moved north. The first continent to separate was Southern Africa, followed by India, Australia and New Zealand, while East Antarctica drifted slowly south, towards the Pole.

Finally, around 30 million years ago, the peninsula broke away from South America, allowing the Southern Ocean to flow freely around the whole continent. This cut it off from more temperate climes and stopped warm air and waters from penetrating south. This isolation led to a significant drop in temperatures on the continent and, in a land where there had once been trees and dinosaurs, the ice, forming first in the centre of the continent, gradually grew until it covered the whole land mass, even flowing out into the ocean as great ice shelves. The land that was once green was entombed beneath kilometres of ice and didn't see the light of day for another 30 million years.

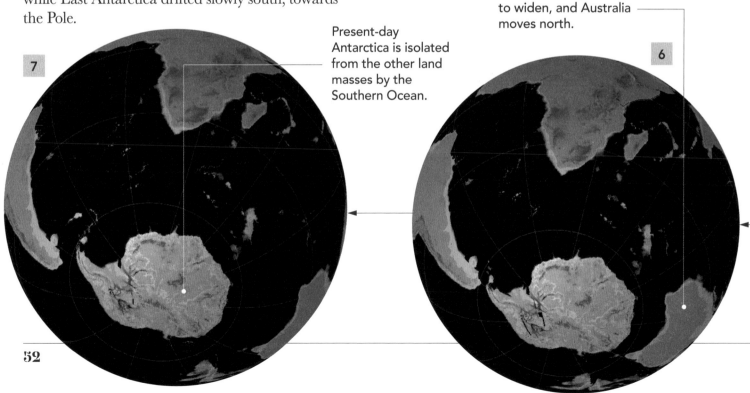

7

Present-day Antarctica is isolated from the other land masses by the Southern Ocean.

30 million years ago, Drake's Passage starts to widen, and Australia moves north.

6

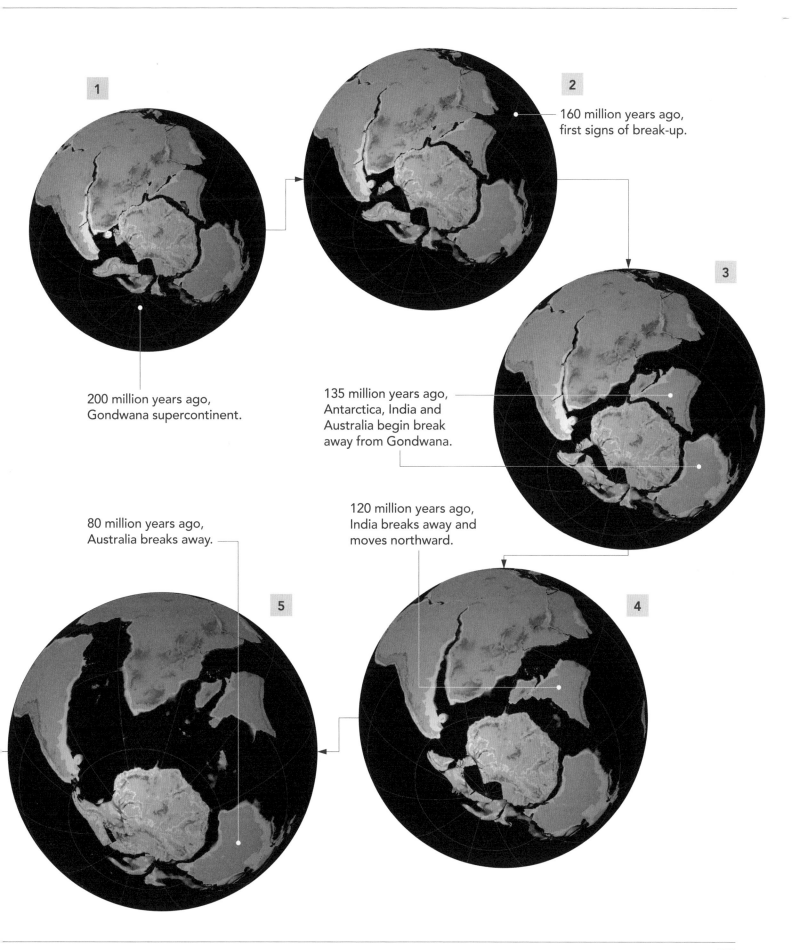

**1** 200 million years ago, Gondwana supercontinent.

**2** 160 million years ago, first signs of break-up.

**3** 135 million years ago, Antarctica, India and Australia begin break away from Gondwana.

**4** 120 million years ago, India breaks away and moves northward.

**5** 80 million years ago, Australia breaks away.

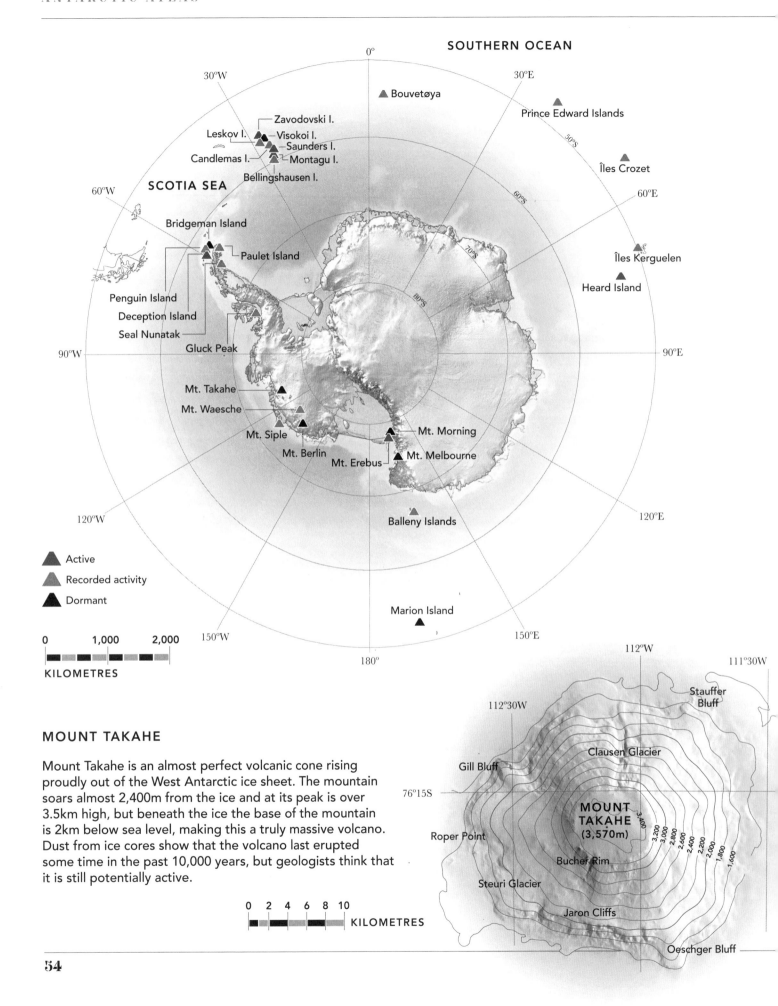

SOUTHERN OCEAN

0°

30°W     30°E

▲ Bouvetøya

▲ Prince Edward Islands

Zavodovski I.

Leskov I.   Visokoi I.

Saunders I.

Candlemas I.   Montagu I.

Bellingshausen I.

50°S

▲ Îles Crozet

SCOTIA SEA

60°W     60°E

60°S

Bridgeman Island

70°S

▲ Îles Kerguelen

Paulet Island

▲ Heard Island

Penguin Island

80°S

Deception Island

Seal Nunatak

90°W     90°E

Gluck Peak

Mt. Takahe

Mt. Waesche

Mt. Morning

Mt. Siple

Mt. Berlin    Mt. Melbourne

Mt. Erebus

120°W     120°E

▲ Balleny Islands

**Active**

**Recorded activity**

**Dormant**

▲ Marion Island

150°W     150°E

180°

0   1,000   2,000

KILOMETRES

## MOUNT TAKAHE

Mount Takahe is an almost perfect volcanic cone rising proudly out of the West Antarctic ice sheet. The mountain soars almost 2,400m from the ice and at its peak is over 3.5km high, but beneath the ice the base of the mountain is 2km below sea level, making this a truly massive volcano. Dust from ice cores show that the volcano last erupted some time in the past 10,000 years, but geologists think that it is still potentially active.

0   2   4   6   8   10    KILOMETRES

112°W

111°30W

Stauffer Bluff

112°30W

Clausen Glacier

Gill Bluff

76°15S

MOUNT TAKAHE (3,570m)

Roper Point

3,400   3,200   3,000   2,800   2,600   2,400   2,200   2,000   1,800   1,600

Bucher Rim

Steuri Glacier

Jaron Cliffs

Oeschger Bluff

# 19. Volcanoes

**ANTARCTICA HAS SOME** of the world's most fascinating volcanoes. Two of the Earth's six lava lakes are here, as well as huge dormant volcanic cones, deceptive hollow calderas and exploded remnant craters. These maps show a selection of some of the continent's most charismatic volcanic mountains, with an overview map (*left*) of the exposed and active volcanoes in or around the Southern Ocean. How many more potentially active mountains lie under the ice sheets remains a mystery. Some scientist believe that parts of West Antarctica have a higher density of volcanoes under the ice than virtually any other place on Earth. When these volcanoes erupt under the ice they can melt the ice above them causing depressions on the surface that can sometimes be seen in satellite imagery.

## MOUNT SIPLE

Mount Siple is a massive stratovolcano that rises steeply from the coast of West Antarctica. Like Takahe (*left*), it is almost perfectly symmetrical, but unlike Takahe it has never been climbed; it is thought to be the world's most prominent unclimbed peak. Although there are no reliable eye-witness accounts of the mountain erupting, ash in nearby ice cores suggests the volcano has been active in the last few thousand years and, although dormant, could still erupt.

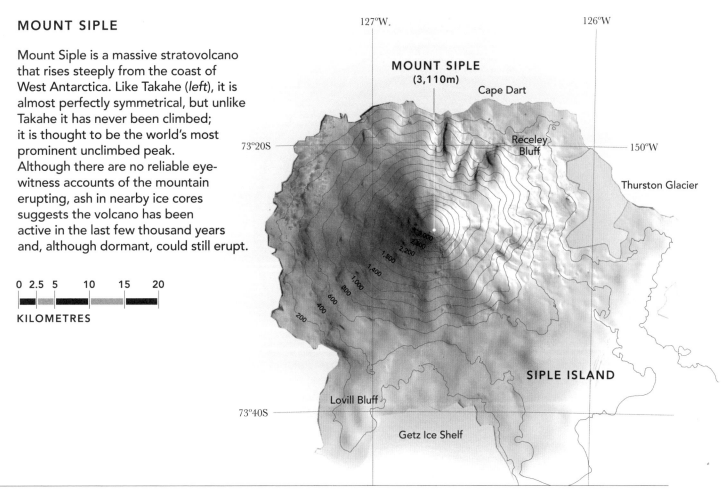

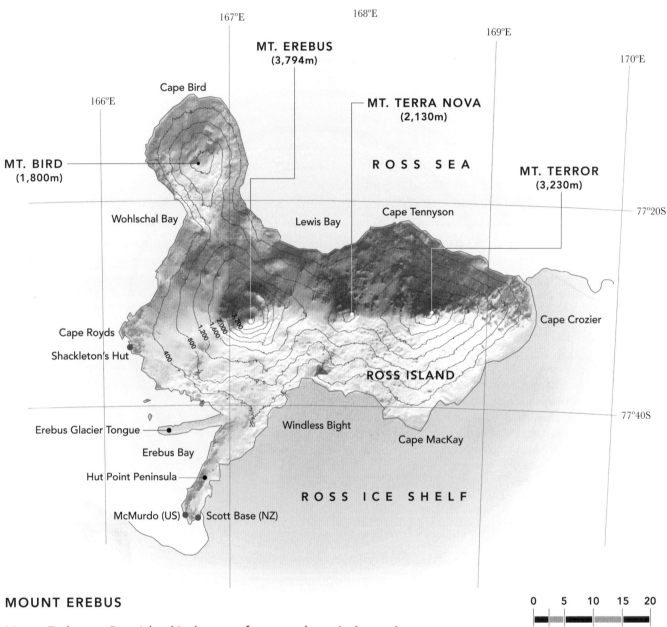

167°E

168°E

169°E

170°E

166°E

**MT. EREBUS**
**(3,794m)**

Cape Bird

**MT. TERRA NOVA**
**(2,130m)**

**MT. BIRD**
**(1,800m)**

R O S S   S E A

**MT. TERROR**
**(3,230m)**

77°20S

Wohlschal Bay

Lewis Bay

Cape Tennyson

Cape Crozier

Cape Royds

1,200

800

3,200

2,000

1,600

400

Shackleton's Hut

**ROSS ISLAND**

Erebus Glacier Tongue

77°40S

Windless Bight

Cape MacKay

Erebus Bay

Hut Point Peninsula

**R O S S   I C E   S H E L F**

McMurdo (US)        Scott Base (NZ)

## MOUNT EREBUS

Mount Erebus on Ross Island is the most famous volcano in Antarctica.
Like Mount Michael (*right*), it has a rare open lava lake that has erupted many
times since it was first discovered by early explorers in the nineteenth century.
It is a huge volcano, rising almost 3,800m straight out of the sea and is one of
four separate volcanic cones on Ross Island. But only Erebus is active.
For many years explorers have used the landmark of the smoking volcano as
a base to explore the interior of the continent. This island is still favoured today,
with two large research bases located at the end of Hut Point Peninsula,
hugging the coast, as far from the wrath of the mountain as possible.

0    5    10    15    20

KILOMETRES

## MOUNT MICHAEL

Mount Michael on Saunders Island lies in the violently tectonic South Sandwich Archipelago, home to at least seven potentially active volcanoes. The island is continually active and has a rare open lava lake, one of only seven in the world. Mount Erebus (*left*) on the other side of the continent holds one of the others, but what makes the Saunders Island lake unique is that it has never been visited by humans; although people have landed on the island, no one has yet managed to climb the steep and angry volcano to peer over its rim into the fiery abyss. The only proof of the lava lake's existence is from satellite imagery; the image that underlies this map shows hot red lava at the base of the volcanic crater.

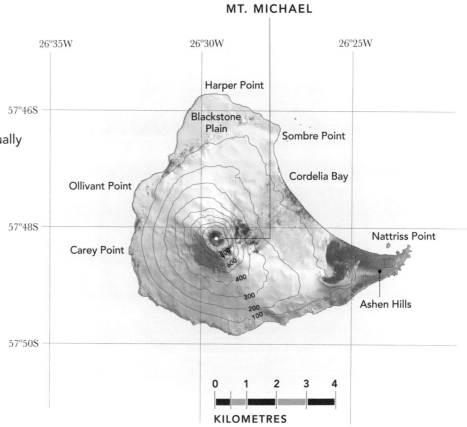

## DECEPTION ISLAND

Deception Island lives up to its name. It looks like an ordinary island from the outside, but pass through the narrow, dangerous channel of Neptune's Bellows and you find yourself within a perfect natural harbour: the collapsed caldera of a huge volcano. The volcano is still active, and its ash buried a British research station in 1969. Today, tour ships often sail into the tranquil waters of the inner bay and tourists wade in the geothermally heated pools.

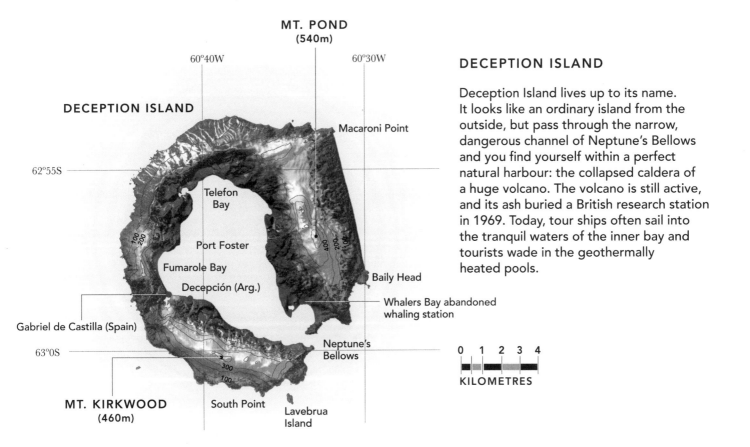

# 20. *The quaking sea*

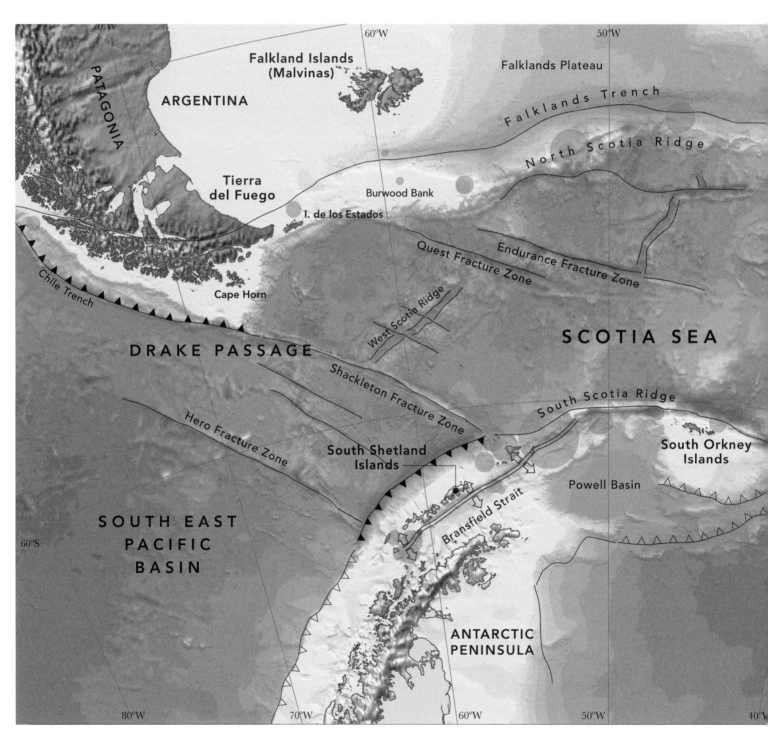

**THE SCOTIA SEA** divides the Atlantic Ocean from Antarctica. It is bounded by South Georgia and the Falkland Islands to the north, the South Shetland and South Orkney Islands to the south, and the South Sandwich Islands to the east. This sea is often extremely rough and exceptionally biologically productive, but what's really interesting is its geology.

Surrounded by moving plate boundaries, the Scotia Sea is one of the most geologically active areas on Earth. It displays every example of tectonic activity possible: a perfect geological laboratory. Over a dozen volcanoes rise from the seabed to make islands, as well as ocean ridges with hydrothermal vents, a deep ocean trench and several major active faults.

One of the main consequences of all this tectonic movement is earthquakes – lots and lots of them.

The map shows the locations of these earthquakes around the periphery of the Scotia Sea, a tectonic zone called the Scotia Arc. Earthquakes characterize active plate margins, and the movement of the South American, Antarctic, Scotia, South Shetland and Sandwich plates is no exception. The greatest concentration of earthquakes occurs near the eastern edge of the Scotia Sea. Here, the ocean floor of the South American plate dives into the Earth's deep interior at the South Sandwich Trench subduction zone, and the small Sandwich Plate, on which the South Sandwich Islands sit, moves eastwards over it. It is here that the largest (indicated by the biggest circles) and deepest (the darker-coloured circles) earthquakes occur as the crust sinks deeper and deeper into the Earth's interior. The release of gases and water causes melting at depth, which creates the volcanoes of the South Sandwich Archipelago, some of the most violently active on Earth.

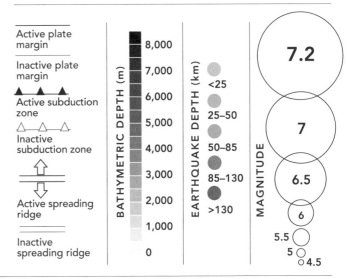

# 21. The driest place on Earth

**THE McMURDO DRY VALLEYS** are considered by many to be the driest place on Earth. It has not rained here for over 3 million years, and the extremely low humidity means that even the rare snow that falls evaporates quickly back into the atmosphere before it gets the chance to melt. The valleys are ice-free, as the mountains at the head of the valleys block the flow of ice from the continental interior, making this the largest area of exposed bedrock in Antarctica. But the extreme cold and hyper-aridity of the valleys make higher life impossible. No organism larger than bacteria survives; the valleys are dead. This otherworldly area has been used by NASA as an analogue for future missions of what conditions on Mars might be like.

Paradoxically, this, the driest place on Earth, is also home to the largest lakes and the longest river in Antarctica. For lakes to exist here, they need to be as extreme as the environment they occupy – they are covered in ice to stop evaporation and are hyper-saline to keep them liquid. Like the valleys, they are lifeless.

On this map, highlighted in orange, are some of the most interesting features of this remarkable place.

1. *The Labyrinth* A deeply eroded series of channels cut into the headland of the Wright Valley. Early explorers battling to navigate this maze of passages likened the area to the mythical Greek labyrinth.

2. *Linnaeus Terrace* In the Dry Valleys, bacteria dominate the ecological niches which other higher organisms usually take. This sunny terrace has one of the richest examples of how these bacterial communities can live within the rocks and the soil.

3. *Victoria Valley sand dunes* Deserts have sand dunes, and the Dry Valleys are no exception. Here, a diverse selection of dunes straddles the valley, the only true sand dunes to occur on the continent.

4. *Boulder Pavement* An area of flat slabs of rock where the Onyx River spreads out, leaving a mosaic of stepping stones across the valley.

5. *Onyx River* A meltwater stream that flows through the Wright Valley. At 32.8 kilometres, it is the longest river in Antarctica. It exists only for a few short weeks each summer, when temperatures in the valley rise slightly above freezing. Strangely, the river flows away from the ocean, from the eastern glaciers flowing into Lake Vanda at the centre of the valley, where the meltwater eventually dries out in the extreme aridity of the area.

6. *Blood Falls* A red, iron-oxide-rich outflowing of saltwater which stains the snout of the Taylor Glacier a deep and disturbing blood-red colour.

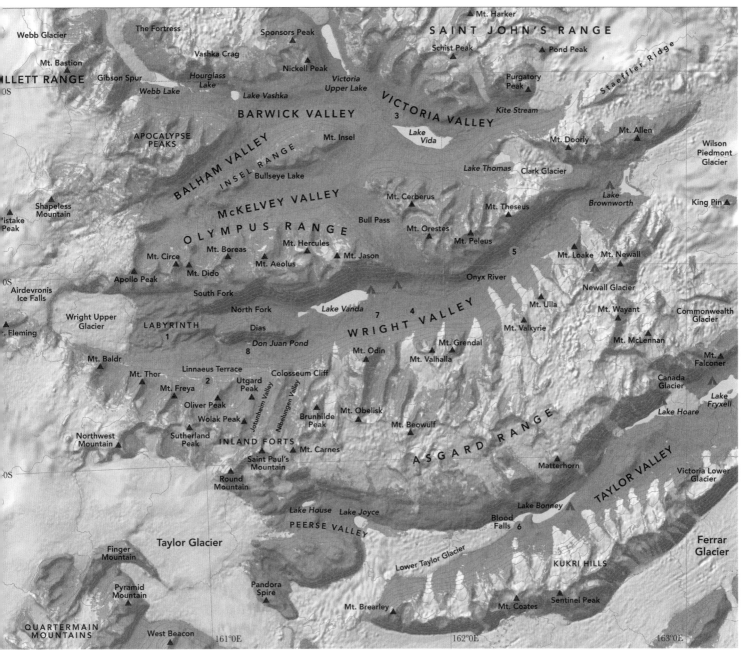

Webb Glacier • The Fortress • Sponsors Peak • SAINT JOHN'S RANGE • Mt. Harker ▲
Schist Peak ▲ • Pond Peak ▲
Mt. Bastion ▲ • Vashka Crag • Nickell Peak ▲ • Purgatory Peak
ILLETT RANGE • Gibson Spur • Hourglass Lake • Victoria Upper Lake • Staeffler Ridge
Webb Lake • Lake Vashka • VICTORIA VALLEY • Kite Stream
3
BARWICK VALLEY • Lake Vida • Mt. Doorly ▲ • Mt. Allen ▲ • Wilson Piedmont Glacier
APOCALYPSE PEAKS • BALHAM VALLEY • Mt. Insel • Lake Thomas • Clark Glacier
INSEL RANGE • Bullseye Lake • Lake Brownworth • King Pin ▲
Shapeless Mountain ▲ • McKELVEY VALLEY • Mt. Cerberus ▲
'istake Peak • OLYMPUS RANGE • Bull Pass • Mt. Theseus ▲
Mt. Orestes ▲ • Mt. Peleus ▲
Airdevronis Ice Falls • Mt. Circe ▲ • Mt. Boreas ▲ • Mt. Hercules ▲ • Mt. Jason ▲ • 5 • Mt. Loake ▲ • Mt. Newall ▲
Mt. Dido ▲ • Mt. Aeolus ▲
Apollo Peak ▲ • Onyx River • Newall Glacier
Wright Upper Glacier • South Fork • Mt. Ulla ▲ • Mt. Wayant ▲ • Commonwealth Glacier
. Fleming • North Fork • Lake Vanda • 7 • 4 • WRIGHT VALLEY
LABYRINTH • Dias • Mt. Valkyrie ▲ • Mt. McLennan ▲
1 • Don Juan Pond • Mt. Odin ▲ • Mt. Grendal ▲
8 • Mt. Valhalla ▲ • Mt. ▲ Falconer
Mt. Baldr ▲ • Linnaeus Terrace • Colosseum Cliff • Canada Glacier
Mt. Thor ▲ • 2 • Utgard Peak • Lake Fryxell
Mt. Freya ▲ • Mt. Obelisk ▲ • Lake Hoare
Oliver Peak ▲ • Brunhilde Peak ▲ • Mt. Beowulf ▲
Wolak Peak ▲ • ASGARD RANGE
Northwest Mountain ▲ • Sutherland Peak • INLAND FORTS • Mt. Carnes ▲ • Matterhorn ▲ • TAYLOR VALLEY • Victoria Lower Glacier
Saint Paul's Mountain • Lake Bonney
Round Mountain • Lake House • Lake Joyce • Blood Falls • 6
PEERSE VALLEY • KUKRI HILLS • Ferrar Glacier
Finger Mountain • Taylor Glacier • Lower Taylor Glacier
Pyramid Mountain ▲ • Pandora Spire • Mt. Brearley ▲ • Mt. Coates ▲ • Sentinel Peak ▲
QUARTERMAIN MOUNTAINS • West Beacon • 161°0E • 162°0E • 163°0E

▲ BASE CAMPS

KILOMETRES
10
8
6
4
2
0

7. *Wright Valley place names* This is the central valley of the group. The mountains on the northern side are named after ancient Greek gods and heroes, while, to the south, the place names reflect Viking and Saxon characters. Here, the gods of Olympus stand guard in an eternal face-off with the gods of Asgard across the valley.

8. *Don Juan Pond* The water in this shallow lake is around 40 per cent salt, making it the most saline lake on Earth. But life still survives; several types of bacteria are known to thrive here.

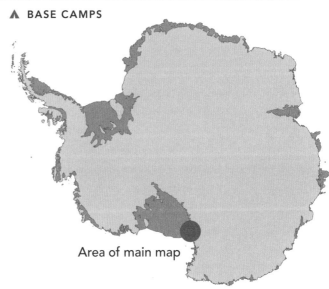

Area of main map

# 22. *Alien invasion*

ISLANDS OFTEN HAVE a disproportionately large percentage of species that are unique to them, a characteristic that ecologists call 'endemism'. Millions of years of isolation, often with few predators, have driven island animals to evolve into many ultra-specialized niches. The more isolated the island, the more specialized and different the animals and plants living on them tend to be.

As the islands of the Southern Ocean are among the most remote in the world, they have a large amount of endemism and many rare and unique species. This is particularly true of birds, especially albatrosses and petrels, which thrive in the vast, tempestuous waters that circle Antarctica.

Endemism requires stability, and the arrival of humans on these islands has been catastrophic. Hunting and habitat loss have taken a massive toll, but perhaps the most critical problem has been the introduction of non-native species to the ecosystems of these islands. Early sailors and settlers brought a wide variety of animals with them, sometimes knowingly, such as pigs and reindeer for food, sometimes inadvertently, such as cats, rats and mice from the ships. This has led to many extinctions and the rapid decline of various other populations, leaving them critically endangered. For the indigenous ground-nesting albatrosses and petrels that evolved on these islands devoid of land-based predators, cats and rats are disastrous and, in many places, their

presence has pushed the birds off the mainland entirely. Today, breeding colonies of many of these native birds are safe only on a few small rocky islets and sea-stacks which the invaders cannot reach.

Over recent years, conservationists and local governments have spearheaded programmes to try to exterminate the invaders. Rats and reindeer have been eradicated from South Georgia. It is hoped that the native birds will re-colonize the areas they once inhabited and that this, along with other conservation efforts, will bring the populations of these unique birds back from the brink of extinction.

The map shows the types of non-native species that have invaded each island in the Southern Ocean (the red boxes), the few remaining pristine islands (green circles), and places where, in some cases, effective eradication projects have removed the aliens (crossed boxes).

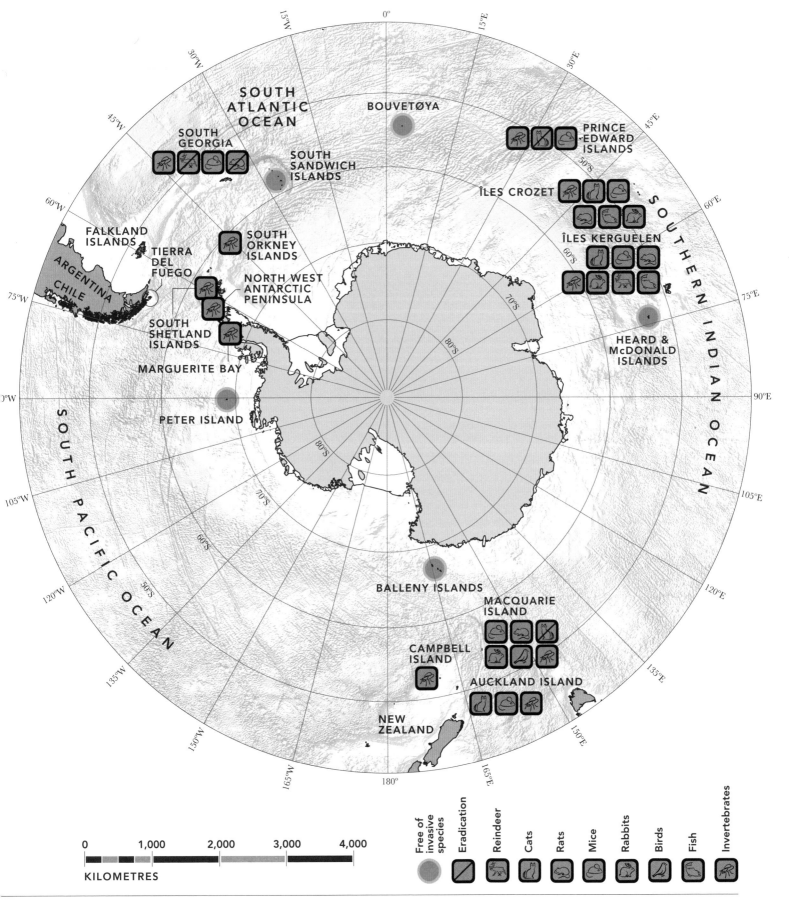

# 23. *Majestic mountains*

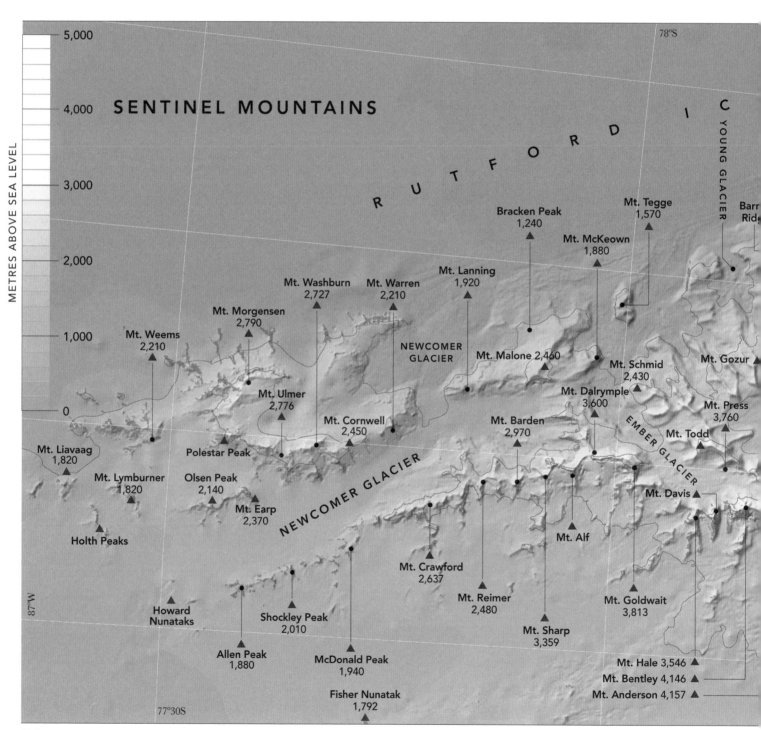

**ALMOST THE ENTIRE** surface of Antarctica is buried beneath the ice; what little rock pokes out above the snow is in the form of mountains. This means that mountains are strangely significant in the frozen landscape – everything other than the ice is mountain.

Many great mountain ranges grace the continent. Dominant among these are the Transantarctic

Mountains, which run from one side of Antarctica to the other, cutting the continent in half. This massive mountain chain, the

*(Continued overleaf)*

KILOMETRES (CONTOUR INTERVALS AT 1,000m)

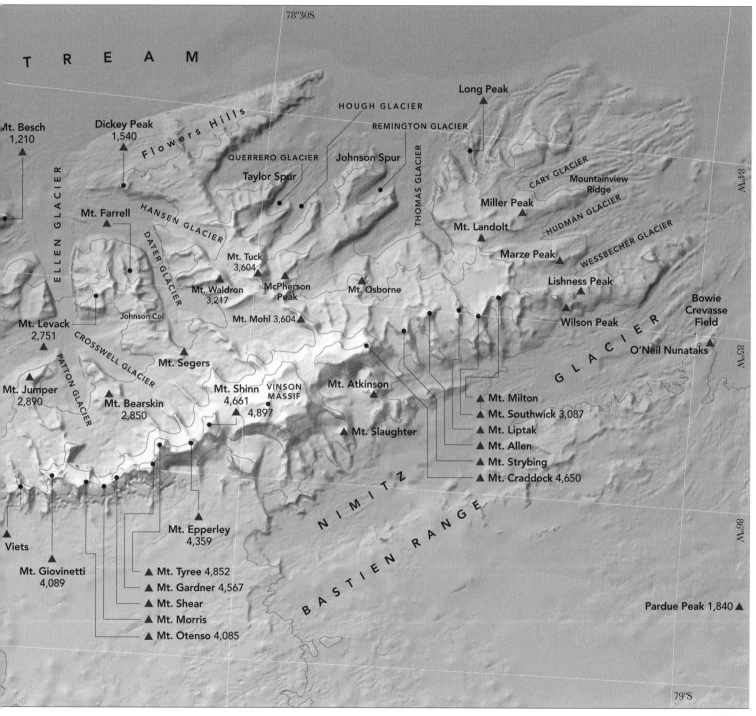

fourth longest in the world, stretches over 3,500 kilometres, from the Argentina Range that borders the Ronne Ice Shelf in the west, south past the Pole and on, up the coast of the Ross Ice Shelf, to the far eastern coast of Antarctica. The highest peak in the range is Mount Kirkpatrick, at 4,528 metres above sea level. It is not, however, the highest mountain on the continent. That honour goes to Vinson Massif in the Ellsworth Mountains. Standing at 4,892 metres, this peak is a Mecca for climbers, many of whom aspire to climb the tallest peak of all seven continents.

The on the previous pages shows the Sentinel Mountains, the main part of the Ellsworth Mountain Range. This range includes Mount Vinson and the four other highest peaks on the continent: Mount Tyree (4,852 metres), Mount Shinn (4,661 metres), Mount Craddock (4,650 metres) and Mount Gardiner (4,567 metres). To the north, the range is bordered by the mighty Rutford ice stream, a deep, fast-flowing glacier over 25 kilometres wide that runs into the sea a few tens of kilometres to the east. Mount Vinson

lies only 40 kilometres from this low-lying ice river, and the high mountains rise sheerly from this low coastal ice to their lofty peaks.

Mount Vinson isn't the hardest mountain to climb; there are many others with steeper and more perilous peaks. In my opinion, the most stunning mountain range in the world is the Drygalski Mountains, part of the Dronning Maud Mountains, a huge curve of mountainous land running near the coast of Dronning Maud Land comprising several prominent ranges. The Drygalski Mountains include a number of impossibly sheer spikes and spires that rise like spears straight from the ice sheet. The most famous – Ulvetanna, or Wolf's Fang – is legendary among climbers and extreme-sports fanatics.

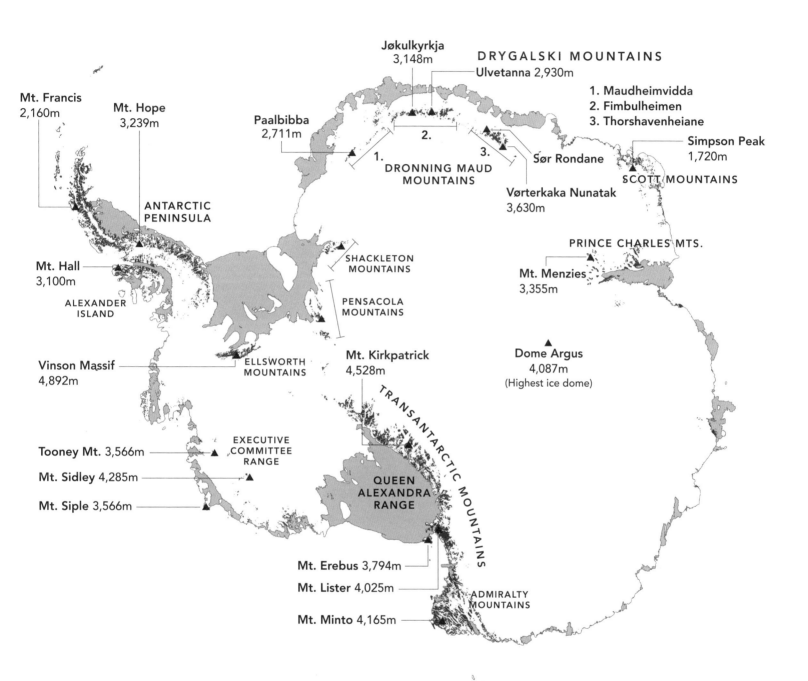

Jøkulkyrkja
3,148m

DRYGALSKI MOUNTAINS

Ulvetanna 2,930m

Mt. Francis
2,160m

Mt. Hope
3,239m

Paalbibba
2,711m

1. Maudheimvidda
2. Fimbulheimen
3. Thorshavenheiane

Simpson Peak
1,720m

2.

3.

1.

Sør Rondane

DRONNING MAUD
MOUNTAINS

SCOTT MOUNTAINS

Vørterkaka Nunatak
3,630m

ANTARCTIC
PENINSULA

Mt. Hall
3,100m

SHACKLETON
MOUNTAINS

PRINCE CHARLES MTS.

Mt. Menzies
3,355m

ALEXANDER
ISLAND

PENSACOLA
MOUNTAINS

Vinson Massif
4,892m

Mt. Kirkpatrick
4,528m

Dome Argus
4,087m
(Highest ice dome)

ELLSWORTH
MOUNTAINS

T R A N S A N T A R C T I C

Tooney Mt. 3,566m

EXECUTIVE
COMMITTEE
RANGE

Mt. Sidley 4,285m

QUEEN
ALEXANDRA
RANGE

Mt. Siple 3,566m

M O U N T A I N S

Mt. Erebus 3,794m

Mt. Lister 4,025m

ADMIRALTY
MOUNTAINS

Mt. Minto 4,165m

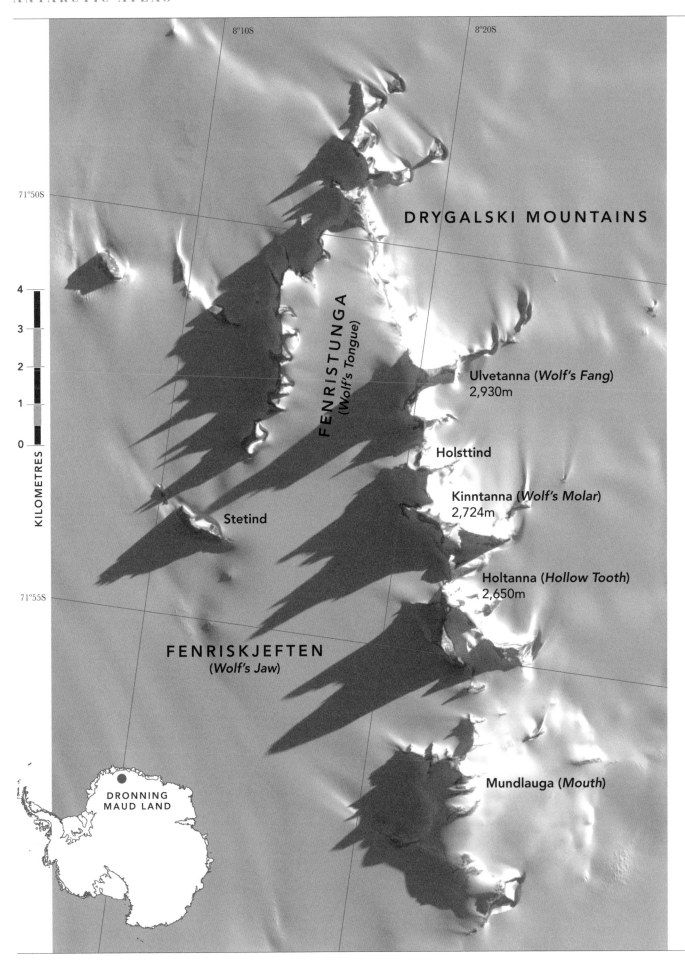

DRYGALSKI MOUNTAINS

FENRISTUNGA
*(Wolf's Tongue)*

Ulvetanna (*Wolf's Fang*)
2,930m

Holsttind

Kinntanna (*Wolf's Molar*)
2,724m

Stetind

Holtanna (*Hollow Tooth*)
2,650m

FENRISKJEFTEN
(*Wolf's Jaw*)

Mundlauga (*Mouth*)

KILOMETRES

4
3
2
1
0

71°50S

71°55S

8°10S

8°20S

DRONNING
MAUD LAND

# 24. The wolf at the end of the world

**HIDDEN UNDER THE ICE**, deep in the Antarctic interior, is the Jaw of Fenris. Fenris, or Fenrir as he is sometimes known, is the monstrous demonic wolf in Norse mythology. The Vikings believed that at Ragnarok, their version of the end of the world, Fenris will wake and wreak havoc on the Earth. The wolf will devour the sun and the moon, then fight and kill Odin, king of the gods, before finally being slain by Víðarr, the Norse god of vengeance.

In East Antarctica's Drygalski Mountain Range lie a group of mountains shaped like a giant wolf's jaw: Fenriskjeften, the 'Wolf's Jaw'. Its spires and sheer spikes rise like teeth from the mouth. These mountains can claim to be possibly the steepest in the world, and the shadows of the peaks on the map give an indication of their jagged shape. Ulvetanna, the 'Wolf's Fang', is the highest in the range at nearly 3 kilometres high. It sits where the canine tooth would be in this monstrous mouth. This mountain has the highest vertical face of any peak on earth, over a kilometre of sheer cliff, straight up out of the ice.

The other peaks are almost as impressive and bear names such as Kinntanna, the 'Wolf's Molar', and Holtanna, 'Hollow Tooth'. The smooth, domed ice cap between the mountains has been named Fenristunga, the 'Wolf's Tongue'.

Such a scenic and spectacular place has not gone unnoticed. Today, the peaks have become a legend among extreme climbers and adrenalin junkies. Ulvetanna was first climbed in 1994. It took sixteen days to ascend the vertical face to the summit, in an epic piece of extreme climbing in sub-zero temperatures and hurricane-strength winds.

# CHAPTER 4

# *Atmosphere*

25. Don't forget your thermals  *72*

26. The hole at the bottom of the world  *74*

27. A future in our hands  *78*

28. Stormy weather  *80*

29. Polar vortex  *82*

AVERAGE TEMPERATURE °C

| 10 | 0 | –10 | –20 | –30 | –40 | –50 |

AVERAGE WINTER WIND SPEED
METRES PER SECOND

| 0.1 | 0.5 | 1.0 | 2.5 | 5.0 | 10.0 |

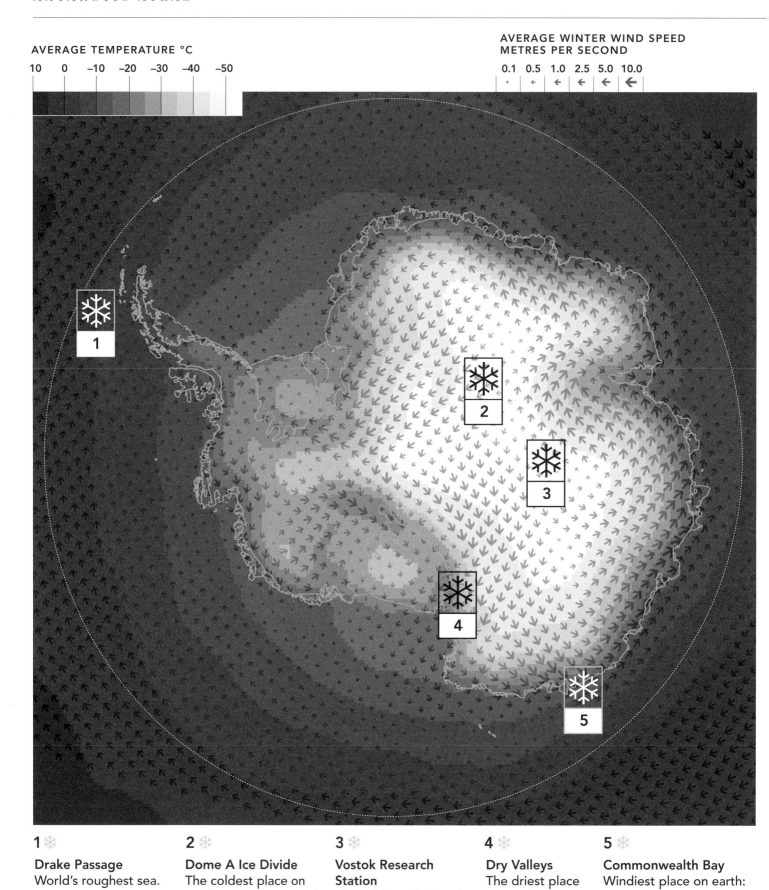

**1** ❄
**Drake Passage**
World's roughest sea.

**2** ❄
**Dome A Ice Divide**
The coldest place on earth: –93°C measured by satellite.

**3** ❄
**Vostok Research Station**
The coldest inhabited place on earth: –89°C.

**4** ❄
**Dry Valleys**
The driest place on earth.

**5** ❄
**Commonwealth Bay**
Windiest place on earth: average wind speed 50mph; maximum wind speed 199mph.

# 25. Don't forget your thermals

**ANTARCTICA IS THE COLDEST**, driest, windiest continent on Earth and is surrounded by the roughest ocean. The temperature on the high polar plateau can get down below -80°C. The record low is at Vostok Research Station: the Russian research base, high on the ice sheet and deep in the interior, once recorded a mind-boggling low of -89.9°C. It's hard to comprehend how cold that really is, but human skin freezes instantaneously below -40°C, so going outside in these conditions is not an option.

During the winter, many areas have exceptionally strong, bitter winds. Streaming off the high ice sheets, cold, dense air sinks towards the coast. In the darkness of the Antarctic winter, with no warmth from the sun, there is nothing to impede the flow of cold air. The freezing winds race downhill, accelerating as the ice sheet gets steeper nearer the coast. By the time they reach sea level, these roaring winds can be extremely strong. At some parts of the East Antarctic coast, the average wind speed is over 50 miles per hour, windier on average than any other place on Earth. Occasionally, in the depths of winter, they can blow at almost 200 miles per hour, stronger than a Class 5 hurricane. No wonder that ill-fated Australian explorer Douglas Mawson, one of the first to overwinter in the region, christened this area the 'home of the blizzard'.

Paradoxically, Antarctica is also the driest continent on Earth. Rain is very rare and restricted to high summer on the northern tip of the warmer Antarctic Peninsula. Most of the interior will never see liquid water, only ice and snow, and there isn't even much snow, with several places limited to less than 2 centimetres a year. Cold air cannot hold much moisture, so the atmosphere itself is very dry – if you go anywhere south of the Antarctic Circle, take some lozenges. In many areas, the high winds sap what little moisture is left in the air. In the Dry Valleys, between the mountains near the Ross Sea, this drying wind has evaporated even what little snow falls here. All that is left is parched, bare rock. Scientists estimate that it has not rained here for 3 million years.

This map shows wind and temperature in Antarctica. Winds are symbolized by arrows. The size of the arrow indicates average winter wind speed, and the direction the arrow points in shows the mean direction. The colour relates to the average yearly temperature at the surface of the ice. The white areas of the high polar plateau are the coldest.

# 26. *The hole at the bottom of the world*

**OZONE IS A NATURALLY** occurring gas that exists in the Earth's atmosphere. It is most plentiful high in the stratosphere, about 20 to 30 kilometres above our heads. This region contains around thirty times more of the gas than other parts of the atmosphere and is commonly termed the 'ozone layer'. Even here, where the ozone is at its densest, it makes up only about ten parts in every million. Though it is rare, the chemical is important for all life on Earth. Ozone shields the world from the sun's harmful ultraviolet (UV) rays that can cause skin cancers and cataracts and can suppress the human immune system. But, above Antarctica, the ozone is missing. There is a hole in the ozone layer.

It has not always been like this. Up until the 1960s, the levels of ozone above Antarctica were stable but, suddenly, in the 1970s, they started to plummet. By the early 1980s, they were down by a third and, at their lowest level, in 1992, the levels were only 28 per cent of what they should have been.

It was scientists at British Antarctic Survey's Halley Research Station who first noticed the problem, in 1984 (although rumour has it that other scientists around the continent had spotted the trend earlier but considered it so odd that they discounted it, thinking that their measuring instruments must be faulty). When the British scientists published their findings, it created a storm: not only was there a hole in the ozone layer over Antarctica, but the size of the hole was growing. If it continued to expand, it would soon engulf countries in the southern hemisphere, such as South America and Australia, causing an increasing health hazard for their populations.

The problem was that, initially, no one knew why the hole had appeared, or what had caused it to grow. After some concerted scientific effort, brilliant thinking and a sprinkle of luck, researchers discovered that a man-made substance – chlorofluorocarbons (CFCs, for short) – was to blame. These gases were commonly used in refrigeration and aerosols, but when they were released into the air they were virtually indestructible, breaking down only when exposed to ultraviolet light. So they diffused upwards in the atmosphere until they were high above the tropics, where the intense UV light broke them down into their basic chemicals. These were then transported polewards until they reached the ozone layer over Antarctica. Here, chlorine acts as a catalyst, destroying the ozone around it. It happened mostly over Antarctica, because only in Antarctica do clouds form in the ozone layer – a consequence of the super-low temperatures – and the chemical reaction that destroys the ozone functions much more effectively with the clouds present.

Once they understood the problem, the international community acted quickly. In one of the most successful pieces of environmental legislation ever brought in, the production of CFCs and other ozone-depleting chemicals was phased out and, ultimately, banned.

It took time for the ozone hole to stop growing. It was at its largest in 2006, but scientists now believe a corner has been turned and the hole is

*(Continued overleaf)*

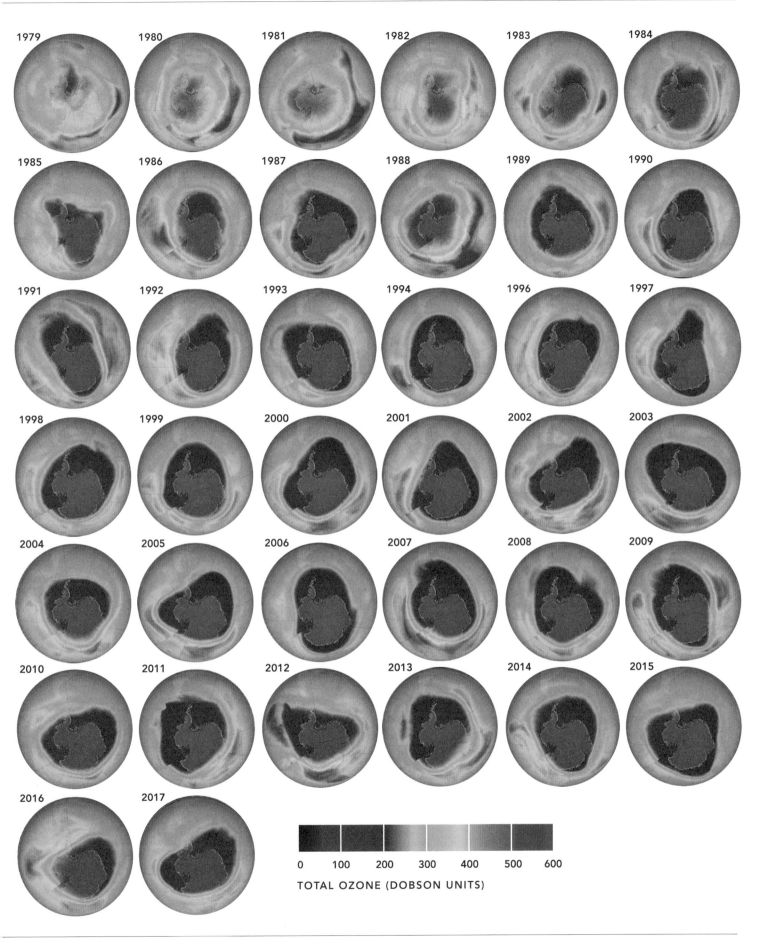

1979 1980 1981 1982 1983 1984
1985 1986 1987 1988 1989 1990
1991 1992 1993 1994 1996 1997
1998 1999 2000 2001 2002 2003
2004 2005 2006 2007 2008 2009
2010 2011 2012 2013 2014 2015
2016 2017

0   100   200   300   400   500   600

TOTAL OZONE (DOBSON UNITS)

slowly starting to shrink. It may take a hundred years to be completely repaired, but this is one environmental disaster that humankind, through scientific effort and political will, has remedied.

The maps on page 75 show the growth and stabilization of the ozone hole since 1979. The blues show low levels of ozone; the greens and reds show greater concentrations.

**THE OZONE HOLE** is a seasonal phenomenon. In February, at the end of the Antarctic summer, ozone levels are relatively stable at around 250–300 DU (Dobson Units, the measure by which ozone is monitored). At this time, ozone values are determined by light from the sun and air traffic in the atmosphere. However, Antarctica is dark during the months of the polar winter and clouds form in the ozone layer. As the sun rises above the horizon once more in the spring (September to November), the UV light interacting with the chlorine from the CFCs rapidly destroys ozone. As spring progresses, the atmosphere warms, the clouds disappear and ozone-rich air once more floods across the continent.

The circular graph shows monthly average levels of ozone for each five-year period since 1970, measured at Halley Research Station. The coloured lines of ozone density are tinted with dark blue for the oldest, through to deep red for the most recent. The record is not complete, as fewer observations have been taken over the harsh winter months when the station is in complete darkness (May to August). The pattern shows that in March, as the polar night approaches, the levels are fairly constant and have not changed much since the 1970s. But in October, in the spring, the levels are very different, with the red and pink lines, the more recent years, extremely low compared to the earlier values. In some periods, there is less than half the amount compared to the early 1970s. It does seem, however, that over the last decade there has been a slight increase in this October nadir, maybe a sign that the global legislation banning CFCs has worked and the hole is possibly showing the first signs of a slow recovery.

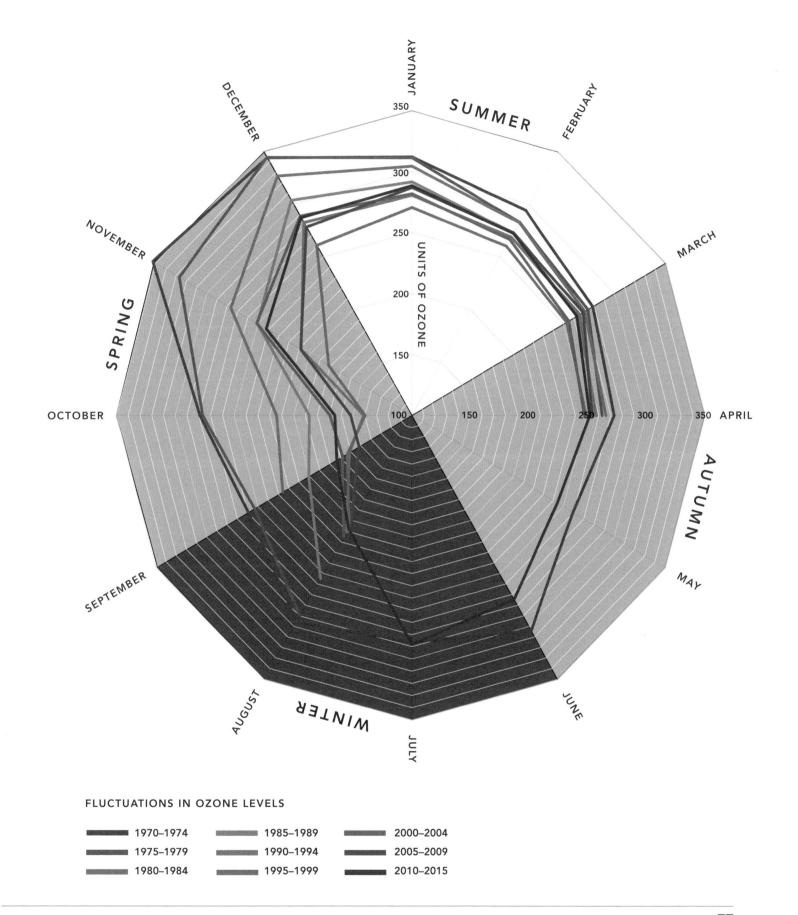

JANUARY

SUMMER

FEBRUARY

DECEMBER

350

300

MARCH

NOVEMBER

UNITS OF OZONE

250

SPRING

200

AUTUMN

OCTOBER

150

100    150    200    250    300    350    APRIL

SEPTEMBER

MAY

AUGUST

WINTER

JUNE

JULY

**FLUCTUATIONS IN OZONE LEVELS**

| | | |
|---|---|---|
| 1970–1974 | 1985–1989 | 2000–2004 |
| 1975–1979 | 1990–1994 | 2005–2009 |
| 1980–1984 | 1995–1999 | 2010–2015 |

LOW-EMISSION
SCENARIO

MODERATE-EMISSION
SCENARIO

HIGH-EMISSION
SCENARIO

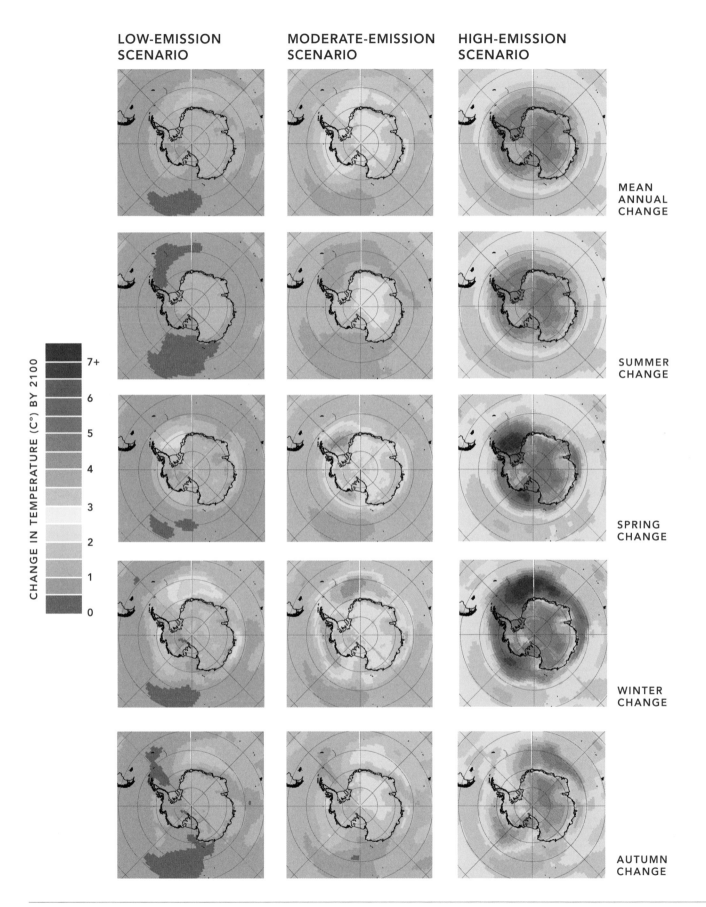

CHANGE IN TEMPERATURE (C°) BY 2100

7+

6

5

4

3

2

1

0

MEAN
ANNUAL
CHANGE

SUMMER
CHANGE

SPRING
CHANGE

WINTER
CHANGE

AUTUMN
CHANGE

# 27. *A future in our hands*

**PREDICTING THE WEATHER** more than a few days in advance is difficult, so predicting what the climate of the Earth will be like in a hundred years' time is highly speculative. As we pump more and more greenhouse gases – for example carbon dioxide and methane – into the atmosphere, we know that the world will change and many parts will become warmer. Understanding these changes – how much warming there will be and where – is crucial for the long-term planning of governments and businesses around the world.

Scientists build multi-parameter models on powerful supercomputers to try to estimate these changes. These models are extremely complex and their accuracy improves every year, but the intricacy of the Earth's system, with its many modes of feedback and nuances, means that there are still some areas where the analysis needs further improvement. One of the most complex regions to understand is Antarctica. The fact that it is so different from the rest of the world often causes global models to characterize this frozen continent poorly. However, the major source of uncertainty in estimates of climate change to 2100 (and beyond) is human actions and predicting how much greenhouse gas we will emit into the atmosphere.

The maps show the very latest estimates from some of these models. They indicate the predicted amount of surface-temperature change in Antarctica by the end of this century. They display annual change and the seasonal changes of summer, autumn, winter and spring. These five predictions are given for three possible scenarios, and each scenario is dependent upon how much carbon dioxide (and other greenhouse gases) humanity pumps into the atmosphere in the future. The first is a 'low-emission scenario', projecting the outcome if we radically reduce our emissions by changing to totally renewable energy sources. The second is a 'moderate-emission scenario', representing a stabilization of climate-change drivers by 2100, which means not emitting more carbon dioxide and other greenhouse gases than we already do. And the third is a 'high-emission scenario: what will happen if we continue to increase our use of greenhouse gases.

The results are quite stark, with up to 4°C of warming by the year 2100 over continental Antarctica in the high-emission scenario. In all scenarios, winter temperatures are set to increase the most. This is due to the predicted loss of winter sea ice. In winter, much of the sea around the continent freezes and the bright white of the ice reflects back much of the sun's warmth. Without the ice, the dark water absorbs much more energy, warming the area around it and increasing temperatures.

But the most striking thing that the maps show is the difference between a future where we curb our emissions and one where we continue to increase them, regardless of the consequences. In essence, it seems that some change is inevitable, but it really is up to us how much the world will change over the next century.

# 28. Stormy weather

**MARINERS FEAR THE SOUTHERN OCEAN**, and for good reason. To sail into the bands of latitude that surround the southernmost seas is something not soon forgotten. These bands – the 'Roaring Forties' and the 'Furious Fifties' – are regions where storms never cease and hurricane-strength winds are always in vogue. The low-pressure systems here are massive and swirl continuously around the globe, unimpeded by land. Strong winds and absence of land mean the waves build up to a mountainous size and give rise to strong currents.

The composite satellite image on the map shows a single day's weather in the southern hemisphere. While the more northerly areas are fairly free of cloud, the seas around Antarctica, especially in the latitudes of 50 to 60 degrees south, have large, swirling areas of cloud – huge low-pressure systems that characterize these latitudes. For humans, the extreme weather means that visiting these areas is challenging, but the wind and waves bring energy to the ocean, and this energy promotes life. The violent mixing of the topmost layers of the ocean increases productivity, and the strong currents driven by the wind cause the upwelling of nutrients from the depths. These nutrients feed the plankton, which, in turn, feed the krill that sustain a rich and abundant Southern Ocean food web supporting the vast numbers of albatrosses, penguins, seals and whales that live in the Southern Ocean.

So, for us the weather might be terrible, but for others it is the very essence of life.

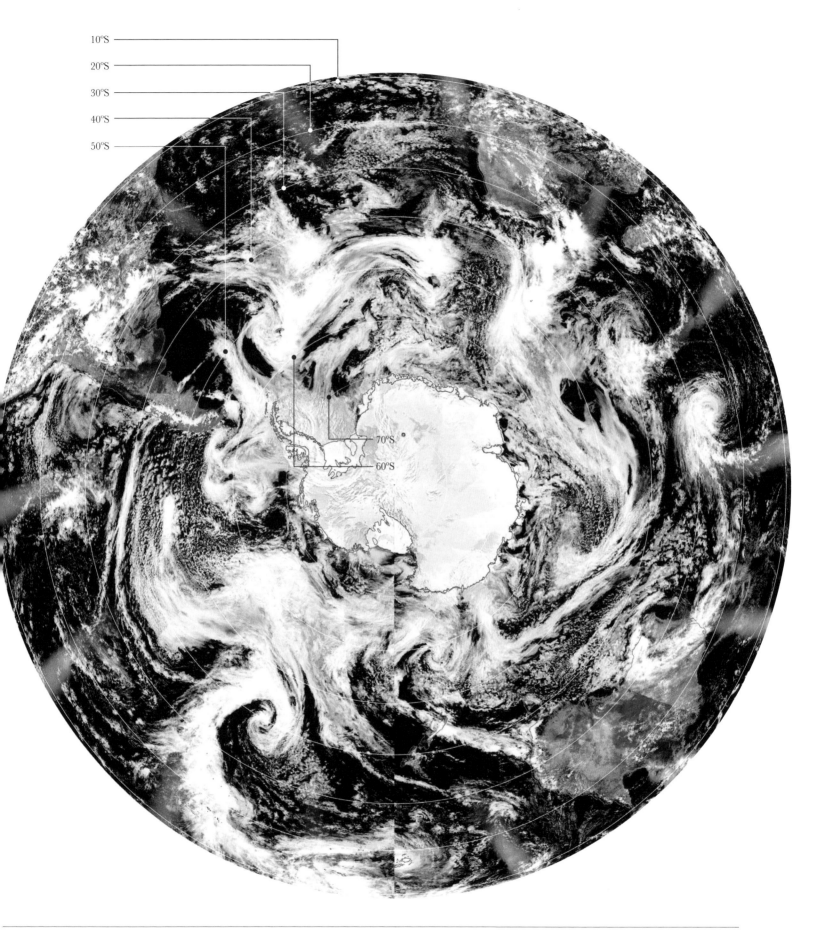

10°S
20°S
30°S
40°S
50°S
60°S
70°S

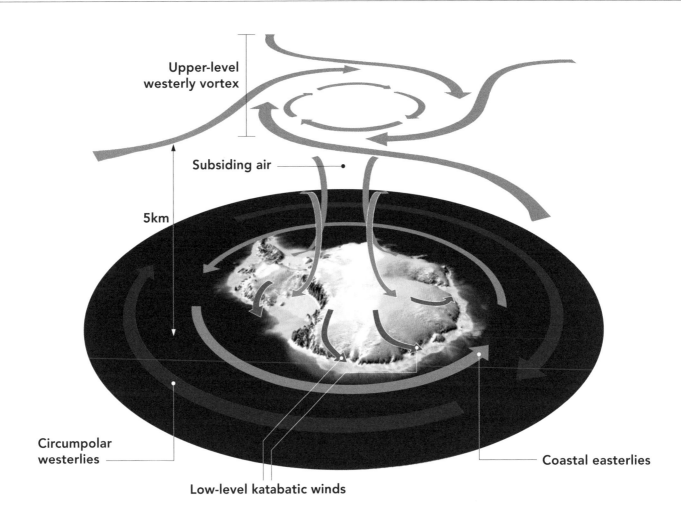

Upper-level
westerly vortex

Subsiding air

5km

Circumpolar
westerlies

Low-level katabatic winds

Coastal easterlies

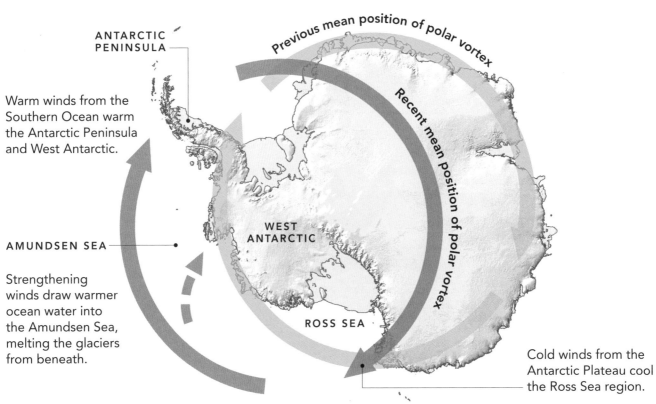

ANTARCTIC
PENINSULA

Previous mean position of polar vortex

Recent mean position of polar vortex

Warm winds from the
Southern Ocean warm
the Antarctic Peninsula
and West Antarctic.

WEST
ANTARCTIC

AMUNDSEN SEA

Strengthening
winds draw warmer
ocean water into
the Amundsen Sea,
melting the glaciers
from beneath.

ROSS SEA

Cold winds from the
Antarctic Plateau cool
the Ross Sea region.

# 29. Polar vortex

**THE POLAR VORTEX** is a strong current of air that circles around Antarctica, high in the upper atmosphere. Its clockwise flow drives the low-pressure systems to go around the Southern Ocean, which, in turn, cuts the continent off from warmer weather systems in the more temperate areas to the north. Within the vortex, the atmosphere is extremely cold, especially in the winter, when the lack of sunlight causes air temperatures to plummet. The extremely cold air in the centre of the vortex descends through the atmosphere to the surface, which is one reason why the surface temperatures of Antarctica are so extreme. Cold air is heavy, so when the super-chilled air mass lands on the ice cap, it flows downhill, like a river, accelerating as the slopes of the ice sheet steepen towards the coast. The descending air creates a vacuum in the upper atmosphere, which pulls stratospheric air southwards from more temperate regions. This in turn will cool and descend, creating a permanent circulation cell that links the tropics and the Poles.

The position and strength of the polar vortex controls much of Antarctica's weather and regional climate. A stronger vortex usually means colder temperatures. Scientists monitoring the vortex have seen it strengthen over the last few decades. They believe that this is due to the development of the ozone hole, which, along with other consequences , led to a cooling of the upper atmosphere, invigorating the cell that drives the circulation. The stronger vortex is probably the reason why we have experienced cooler temperatures over some parts of the continent and why Antarctica has not seen the rapid climate change we have witnessed in the Arctic. Worryingly, this means that, as the ozone hole reduces, the vortex will weaken and temperatures will rise rapidly. Other scientists have pointed to changes in the position of the vortex. Although very variable, it seems that its mean position has shifted away from the centre of the Pole, towards West Antarctica. This shift has strengthened the northward flow of air into the Ross Sea, cooling that region, as super-cold air flows downwards from the Antarctic plateau and out to the coast. Meanwhile, to the west, the Antarctic Peninsula has seen increasing wind speeds, leading to higher temperatures and greater snowfall, while the changing wind direction has drawn warm water towards the West Antarctic coast, melting the large glaciers of the area.

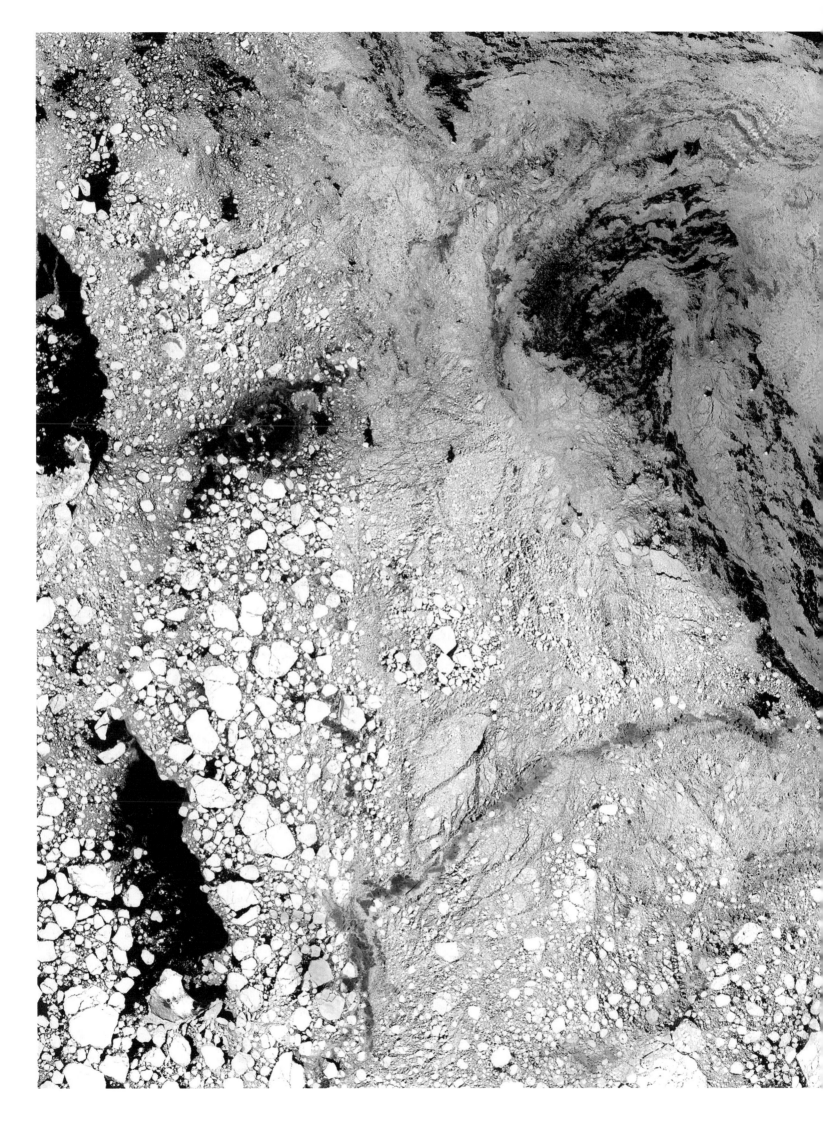

# CHAPTER 5

## *Sea*

30. The Southern Ocean *86*

31. Islands in the stream *88*

32. Ocean currents *92*

33. Ocean eddies *94*

34. The greatest change on Earth *96*

35. The engine of the ocean *98*

36. The life of a berg *100*

37. The green ocean *102*

38. Earth's lungs *104*

# 30. The Southern Ocean

**THE SOUTHERN OCEAN** surrounds Antarctica, separating it from the warmer waters to the north. It is the fourth largest of the world's oceans, after the Pacific, Atlantic and Indian. Defining the exact limits of the Southern Ocean is tricky, and different groups of people use different ways. Politicians look upon the 60-degree line of latitude as the dividing line, while ocean scientists often consider the boundary to be the region where cold Antarctic waters meet warmer waters in the north. This is a narrow frontal zone of rapidly changing temperature typically called the 'polar front' (marked as a thick blue line on the map).

The shades of blue on the map show bathymetry – the depth of the seabed. The lighter blue illustrates shallow waters and the dark blue deep ocean abysses and trenches. The lighter-coloured seabed ridges that circle the continent follow tectonic-plate boundaries, while paler dots denote under-sea mountains called 'seamounts', plateaus and islands.

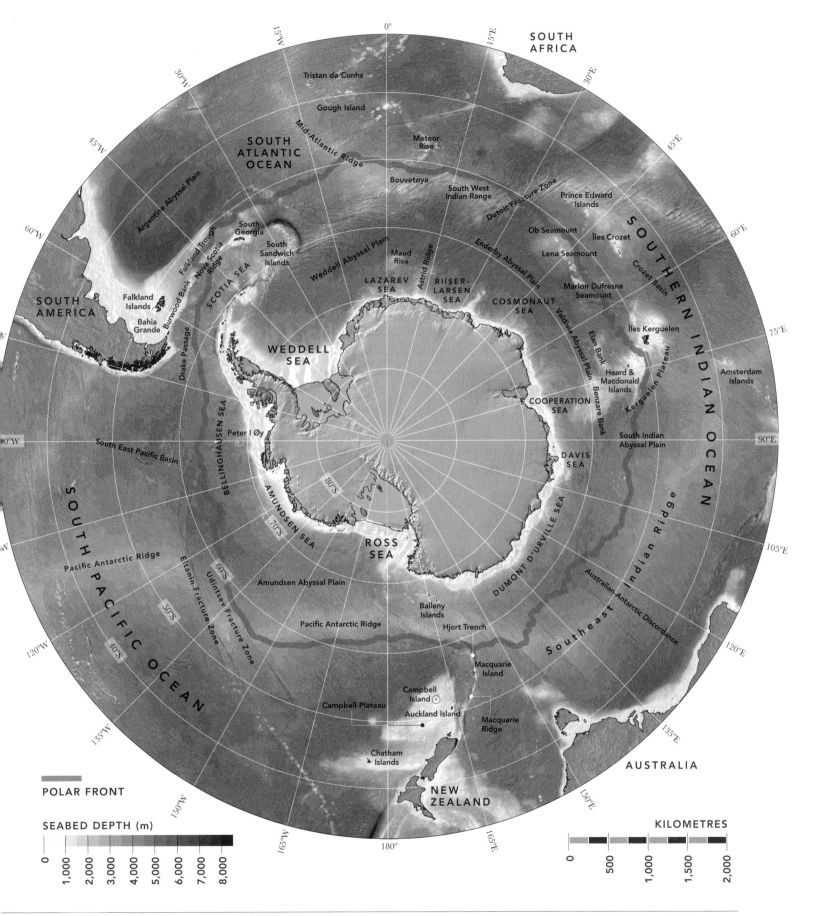

SOUTH AFRICA

SOUTH ATLANTIC OCEAN

Tristan da Cunha

Gough Island

Meteor Rise

Mid-Atlantic Ridge

Bouvetøya

South West Indian Range

Dutoit Fracture Zone

Prince Edward Islands

Ob Seamount

Îles Crozet

Lena Seamount

Argentine Abyssal Plain

South Georgia

Weddell Abyssal Plain

Maud Rise

LAZAREV SEA

Astrid Ridge

RIISER-LARSEN SEA

Enderby Abyssal Plain

Crozet Basin

SOUTHERN INDIAN OCEAN

Marion Dufresne Seamount

Falkland Trough

Nova Scotia Ridge

South Sandwich Islands

COSMONAUT SEA

Valdivia Abyssal Plain

Îles Kerguelen

Burwood Bank

SCOTIA SEA

Elan Bank

Kerguelen Plateau

Amsterdam Islands

SOUTH AMERICA

Falkland Islands

Bahia Grande

Heard & Macdonald Islands

Banzare Bank

Drake Passage

WEDDELL SEA

COOPERATION SEA

South Indian Abyssal Plain

South East Pacific Basin

BELLINGHAUSEN SEA

Peter I Øy

80°S

DAVIS SEA

90°E

SOUTH PACIFIC OCEAN

AMUNDSEN SEA

70°S

ROSS SEA

DUMONT D'URVILLE SEA

Southeast Indian Ridge

Pacific Antarctic Ridge

Eltanin Fracture Zone

Udintsev Fracture Zone

60°S

50°S

Amundsen Abyssal Plain

Pacific Antarctic Ridge

Balleny Islands

Hjort Trench

Australian-Antarctic Discordance

40°S

Macquarie Island

120°E

Campbell Island

Campbell Plateau

Auckland Island

Macquarie Ridge

AUSTRALIA

Chatham Islands

135°E

POLAR FRONT

NEW ZEALAND

150°E

SEABED DEPTH (m)

0 1,000 2,000 3,000 4,000 5,000 6,000 7,000 8,000

KILOMETRES

0 500 1,000 1,500 2,000

180°

165°E

# 31. Islands in the stream

**THE SOUTHERN OCEAN** is a vast, open, bleak sea stretching between the warmer, populated continents to the north and the frigid, desolate coasts of Antarctica to the south. In the middle of this ocean, strung like a string of rare pearls most of the way around the world, are a dozen or so remote, lonely islands. These islands make up some of the most isolated places on Earth, thousands of kilometres from each other and the rest of the world. Lashed by continual winds from the west and by the currents that flow around the Southern Ocean, the islands are bleak, cold places for people to live on and, like Antarctica, none of them had native human populations. Today, just a few researchers and government officials inhabit them. None of the islands are particularly big, but the two larger ones, South Georgia and Kerguelen, are both mountainous, ancient pieces of land that were separated from their continental plates aeons ago, in the distant geological past. Over time, they have wandered, pushed by the power of plate tectonics, out into the ocean, far from other shores. Almost all the smaller islands are volcanoes; they reach up from the deep waters of the ocean to emerge cold and barren above the waves. Some are extinct or dormant, but a number are still active. In the cold climate of the south, many of the islands have glaciers and are snow-covered for much of the year.

The one thing that all these islands share is a wealth of wildlife. In such a lonely ocean, dry land for animals to breed on is at a premium. The islands are dense with penguins, flying seabirds and seals. Although the land mass of these isolated specks is only tiny, there are more penguins breeding on them than on the whole Antarctic mainland. The isolated nature of the land has given birth to many different species of birds, some restricted to a single island, some spread over just a handful of them. Most of the world's albatross species breed on these islands, as well as the majority of the world's penguin species and a number of other rare and endangered seabirds. The other thing that draws wildlife here is the productivity of the surrounding seas. Most of the open ocean is poor in food, but where the strong currents of the Southern Ocean hit the steep undersea slopes surrounding the islands deep water is raised to the surface, bringing nutrients and food. This powers a super-productive food chain on which the rich diversity of birds, seals and whales depend.

## 1. SOUTH GEORGIA (UK)

1. Annenkov Island
2. King Edward Point
3. Allardyce Range
4. Salvesen Range
5. Bird Island

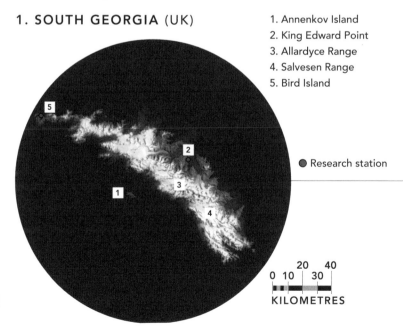

● Research station

20      40
0  10    30
KILOMETRES

Humans have a chequered history here. Most of the islands were discovered during the last two hundred and fifty years, and since then we have pillaged their natural resources until almost nothing remains. It was here that the great whaling factories stripped the seas of whales. Seals were hunted in their millions, almost to extinction, and species introduced from other parts of the world annihilated many of the native birds in much of their breeding habitat. However, in the latter part of the twentieth century, we stopped exploiting and started to conserve these rich and unique places, and for the most part the wildlife has responded. Seals have returned in their millions, whale numbers are starting to recover and, with the carefully managed eradication of some of the non-native species, the flying seabirds are beginning to show signs of recovery. There is still a long way to go, but with time, effort and care, these islands are once more becoming the pearls in the crown of the Antarctic.

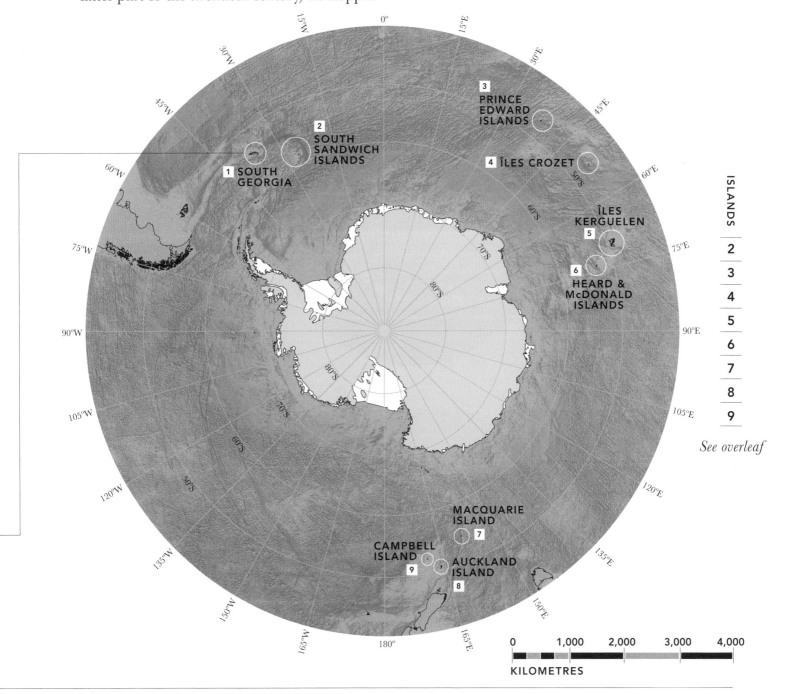

**ISLANDS**

| 2 |
| 3 |
| 4 |
| 5 |
| 6 |
| 7 |
| 8 |
| 9 |

*See overleaf*

0   1,000   2,000   3,000   4,000
KILOMETRES

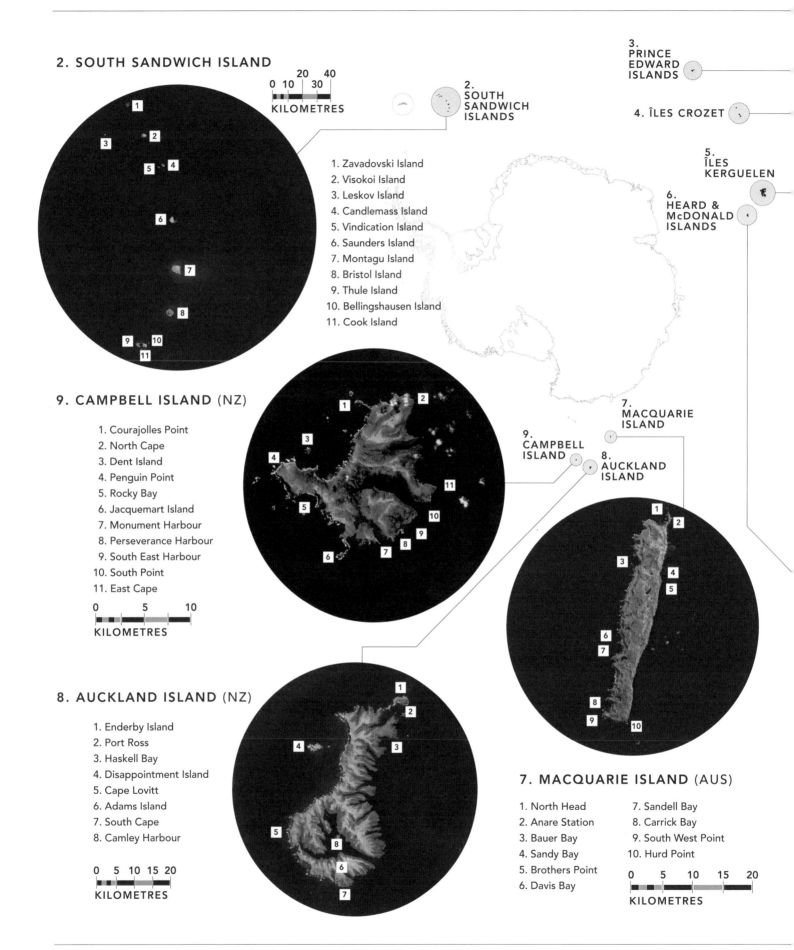

## 2. SOUTH SANDWICH ISLAND

**KILOMETRES** 0 10 20 30 40

1. Zavadovski Island
2. Visokoi Island
3. Leskov Island
4. Candlemass Island
5. Vindication Island
6. Saunders Island
7. Montagu Island
8. Bristol Island
9. Thule Island
10. Bellingshausen Island
11. Cook Island

3. PRINCE EDWARD ISLANDS

2. SOUTH SANDWICH ISLANDS

4. ÎLES CROZET

5. ÎLES KERGUELEN

6. HEARD & McDONALD ISLANDS

7. MACQUARIE ISLAND

9. CAMPBELL ISLAND

8. AUCKLAND ISLAND

## 9. CAMPBELL ISLAND (NZ)

1. Courajolles Point
2. North Cape
3. Dent Island
4. Penguin Point
5. Rocky Bay
6. Jacquemart Island
7. Monument Harbour
8. Perseverance Harbour
9. South East Harbour
10. South Point
11. East Cape

0 5 10 **KILOMETRES**

## 8. AUCKLAND ISLAND (NZ)

1. Enderby Island
2. Port Ross
3. Haskell Bay
4. Disappointment Island
5. Cape Lovitt
6. Adams Island
7. South Cape
8. Camley Harbour

0 5 10 15 20 **KILOMETRES**

## 7. MACQUARIE ISLAND (AUS)

1. North Head
2. Anare Station
3. Bauer Bay
4. Sandy Bay
5. Brothers Point
6. Davis Bay
7. Sandell Bay
8. Carrick Bay
9. South West Point
10. Hurd Point

0 5 10 15 20 **KILOMETRES**

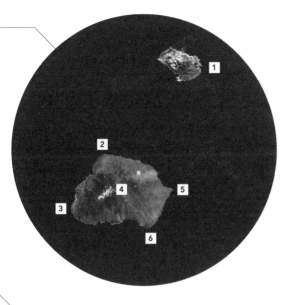

## 3. PRINCE EDWARD ISLAND (SA)

1. Prince Edward Island
2. Cape Davis
3. Cape Crozier
4. Marion Island
5. East Cape
6. Cape Hooker

0 5 10 20 30
KILOMETRES

## 4. ÎLES CROZET (FRANCE)

1. Île des Apôtres
2. Île aux Cochons
3. Île des Pingouins
4. Île de la Possession
5. Île de l'Est

0 5 10 20 30
KILOMETRES

## 5. ÎLES KERGUELEN (FRANCE)

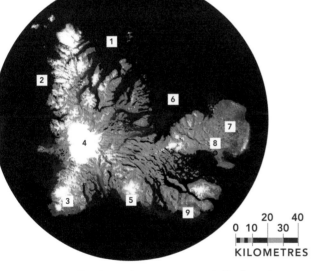

20 40
0 10 30
KILOMETRES

1. Golfe Choiseul
2. Péninsule Loranchet
3. Péninsule Rallier du Baty
4. Calotte Glaciaire Cook
5. Péninsule Gallieni
6. Golfe des Baleiniers
7. Péninsule Courbet
8. Port-aux-Français
9. Péninsule Jeanne d'Arc

## 6. McDONALD ISLAND (AUS)

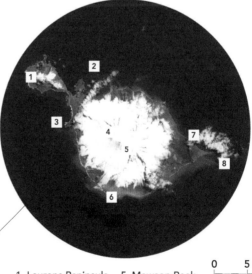

1. Laurens Peninsula
2. Corinthian Bay
3. Cape Gazert
4. Big Ben
5. Mawson Peak
6. Cape Labuan
7. Spit Bay
8. Spit Point

0 5 10
KILOMETRES

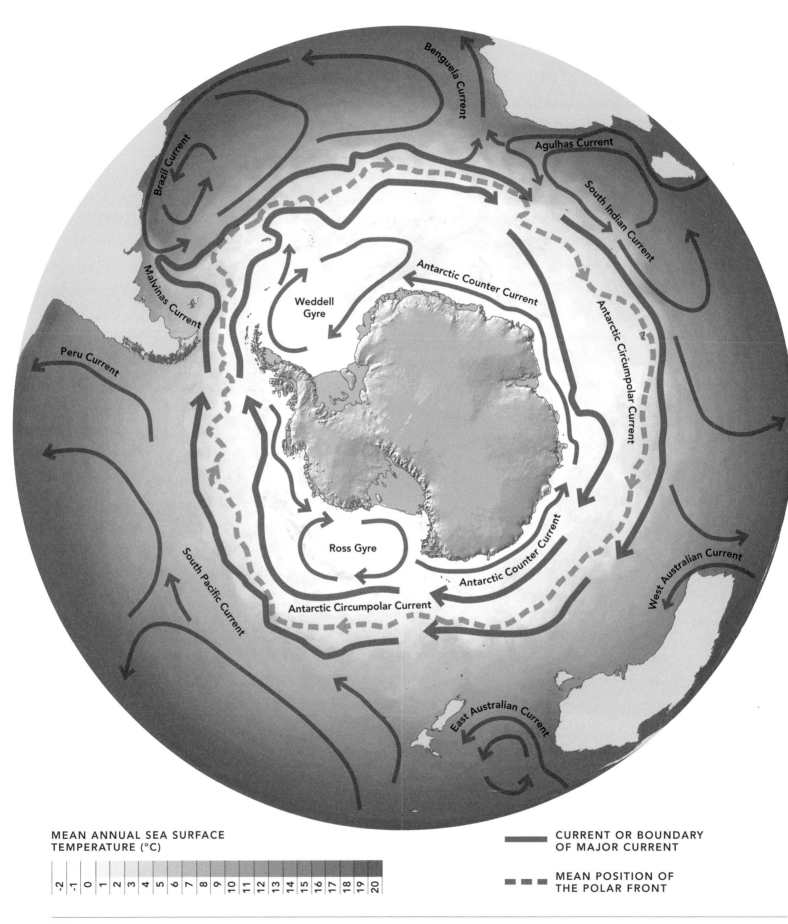

Benguela Current

Brazil Current

Agulhas Current

South Indian Current

Malvinas Current

Antarctic Counter Current

Weddell Gyre

Antarctic Circumpolar Current

Peru Current

Ross Gyre

Antarctic Counter Current

South Pacific Current

Antarctic Circumpolar Current

West Australian Current

East Australian Current

MEAN ANNUAL SEA SURFACE
TEMPERATURE (°C)

| -2 | -1 | 0 | 1 | 2 | 3 | 4 | 5 | 6 | 7 | 8 | 9 | 10 | 11 | 12 | 13 | 14 | 15 | 16 | 17 | 18 | 19 | 20 |
|---|---|---|---|---|---|---|---|---|---|---|---|---|---|---|---|---|---|---|---|---|---|---|

CURRENT OR BOUNDARY
OF MAJOR CURRENT

MEAN POSITION OF
THE POLAR FRONT

# 32. *Ocean currents*

**THE SOUTHERN OCEAN** is characterized by some of the world's strongest currents. Akin to rivers of moving water within the sea, these currents are driven by surface winds, heated by the sun and fresh water from ice melt and precipitation. Unlike anywhere else on the globe, around Antarctica, at a latitude of between 55 and 62 degrees south this flow is unbroken, and the currents move water continuously around the world, flowing without the interruption of land, which can divert and weaken other currents.

This Antarctic Circumpolar Current (ACC) is the strongest feature in the Southern Ocean and the largest system of currents on Earth, with a flow four times greater than the Gulf Stream. Wide and complex, its exact boundary is hard to define, as the current twists into contorted strands and filaments. The overall shape is roughly circular, although there are a number of diversions caused by seamounts (small islands and plateaus rising from the seabed). This continuous flow of water acts as a wall to the warmer waters of the tropics, stopping them penetrating south and reaching the icy coasts of Antarctica. Indeed, it is thought that the vast Antarctic ice sheets were born only when the Drake Passage opened and allowed the circumpolar current to form and the ACC in turn cut Antarctica off from warmer waters further north.

Within this current, the distinction between the warm northern waters and the cold Antarctic water can be surprisingly sharp. Water temperatures can change by 6°C over just a few kilometres. This sharp divide is called the 'polar front' and, like the boundary of the currents, its exact location migrates in complex twists and loops. On this map, the mean location of the polar front is shown by the dashed line.

As well as the Antarctic Circumpolar Current, there are other important features in the Southern Ocean's circulation. To the south of the polar front, close to Antarctica, there is an opposing westward flow of seawater called the Antarctic Coastal Current. This anticlockwise flow hugs the coast and is important for transporting icebergs around Antarctica. There are also two large indentations breaking the circular shape of the continent. These are the Ross Sea and the Weddell Sea, and they both have their own circular currents, or gyres, which continually rotate the sea ice within them.

# 33. Ocean eddies

**THE ANTARCTIC CIRCUMPOLAR CURRENT** (ACC) circles Antarctica, perpetually moving water west to east, but within this current there is great complexity. Eddies, gyres and vortices swirl and merge, driven by contortions and pulses in the speed of the current flow. This map gives an example of that complexity over a single 24-hour period. The background image shows the water movements in terms of eddy kinetic energy of the surface current flow. The most energetic, strongest currents are shown in green, with lighter blue denoting weaker currents. The eddy energy in the ocean follows the same path as the ACC and shows strong connections with the complex regions of ocean current flow next to the major continents of South America, Africa and Australia. As the winds over the Southern Ocean have strengthened in recent decades, the energy contained within the ocean eddies has also increased. The Southern Ocean absorbs much of the $CO_2$ produced by humanity and it has been suggested that the increased wind speed has helped this process. But the sea can only absorb so much and, once saturated, it may not be able to cope with the rising levels of greenhouse gases as quickly as it does now, which will hasten their increase. What the impact of further changes will be on the Southern Ocean and its interaction with the rest of the world is currently a topic of concern within the scientific community.

2000

320

45

7

0

# 34. The greatest change on Earth

**EVERY YEAR**, as the summer draws to a close, as the days get shorter and temperatures start to plummet, the seas around the coast begin to freeze. All around Antarctica, the ocean becomes covered in sea ice. Even in the coldest temperatures, only the surface layer freezes, and even then only ever to a few metres deep. The ice itself insulates the lower layers from the freezing cold, but the area covered by the sea ice is enormous. By the end of the southern winter, an area of the Southern Ocean roughly equal to the size of the continent freezes and the continent literally doubles in size. In the spring, as the sun returns and temperatures begin to rise, almost all of this surface ice melts, until only a small percentage is left by the end of the summer. This seasonal change is perhaps the greatest annual environmental change on Earth.

Strangely, the sea ice in the more northerly parts is not formed in situ; the air temperatures there are rarely cold enough to freeze the sea. Instead, it is created further south and pushed northwards by the constant formation of new ice and the winds. It is near the coast that most of the sea ice is formed, when strong winter winds thunder off the high ice sheets. At the coast, these offshore winds blow the sea ice away from Antarctica, pushing it northwards; at the same time, they create areas of open water near the coastline within which the unfrozen seawater is never less than a few degrees below zero. However, the air temperature here in midwinter can be 50 degrees below zero, so as soon as the wind drops slightly the surface waters refreeze. This massive difference in temperature can lead to some strange weather phenomena – the waters appear to smoke and the surface can look greasy; it then begins to crystallize as the sea ice starts to form. Within a few hours, the sea can freeze over completely and, over the space of a day or two, it will usually solidify into unbroken, solid sea ice, strong enough to hold a person's weight. The next time the wind picks up, this new solid layer will be blown northwards, away from the coast, pushing out the ice in front of it. Then a new lead (channel of water) will open up and the process will begin again. These coastal areas are often known as 'sea ice factories', or 'polynyas', and some of the most important are marked as swirls on the map.

| SEA ICE CONCENTRATION (m) | FEBRUARY | SEPTEMBER |
|---|---|---|
| | 1–10 | 1–10 |
| | 11–20 | 11–20 |
| | 21–30 | 21–30 |
| | 31–40 | 31–40 |
| | 41–50 | 41–50 |
| | 51–60 | 51–60 |
| | 61–70 | 61–70 |
| | 71–80 | 71–80 |
| | 81–90 | 81–90 |
| | 91–100 | 91–100 |

DECEMBER

NOVEMBER

OCTOBER

SEPTEMBER

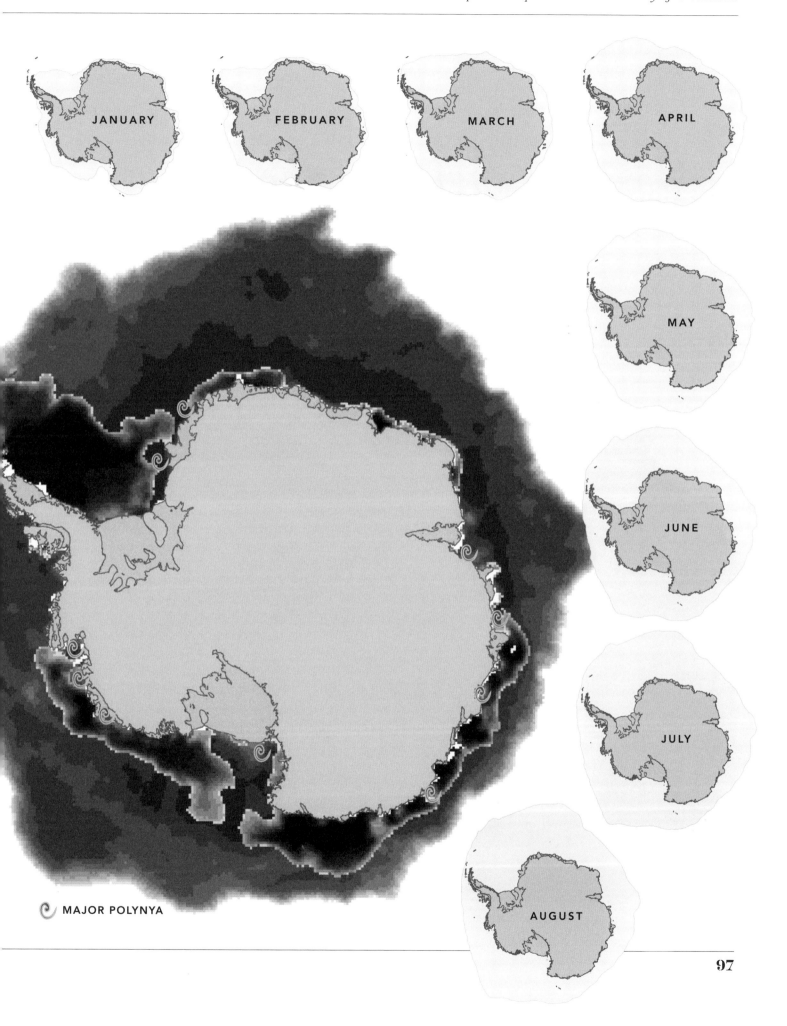

JANUARY

FEBRUARY

MARCH

APRIL

MAY

JUNE

JULY

AUGUST

🌀 **MAJOR POLYNYA**

# 35. The engine of the ocean

**THE EARTH IS A** complex place. Often, processes in one part of the world have a major impact on seemingly unconnected events thousands of kilometres away. One strange connection is the relationship between the deepest regions of the world's oceans and Antarctic sea ice. The replenishment of the deep water in the world's oceans relies heavily on the freezing of the Southern Ocean each Antarctic winter. The resulting spread of the dense water formed in Antarctica affects the deep seas and shapes the climate around the world.

In the deep ocean there is usuall little direct interaction between the surface layers and the water thousands of metres below. However, this is not true in Antarctica. Here, new deep water is being constantly created. Every year, as the sea freezes around the continent, saltwater turns to ice, but this ice contains relatively little salt. The excess salt is expelled into the surrounding sea, often making it extremely saline. Continual refreezing of the surface layer makes the layer just under the surface more and more salty and, as salty water is denser and heavier than fresher water, the water sinks. As the cold, heavy surface water drops into the depths, it pushes the deep ocean water below it northwards. Some of the salty water will flow out at intermediate depths, but much will continue to sink until it reaches the abyssal plains on the ocean floor. Meanwhile, at the surface, warmer, fresher waters from the north will flow in to replace the sinking salty brine. In turn, this will become salty and also sink, creating a constant, never-ending circulation. The deep water flows across the abyssal plains of much of the Earth before slowly, over many decades or centuries, it reaches the surface once more. This process is called 'thermohaline circulation' ('thermo', as the sinking waters are cold, and 'haline', as they are salty); it is the engine of the ocean.

However, if the amount of sea ice that forms each year around Antarctica reduces, as current climate models predict, this engine will slow and the flow of deep water around the world's oceans will be affected. The change could have serious repercussions for weather patterns all over the globe, but exactly what these will be is currently unknown. It is inevitable that changing the Earth's climate will have unforeseen consequences.

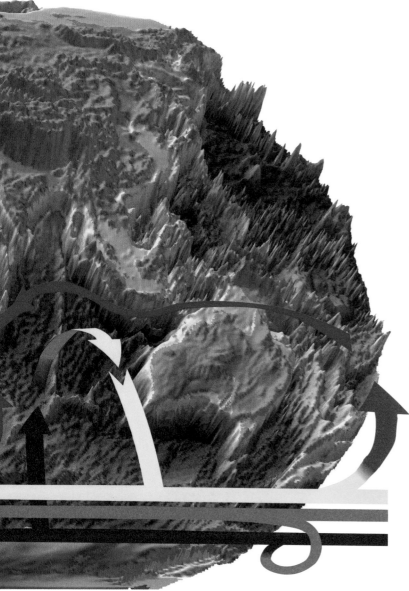

COLD UPPER WATERS

WARMER WATERS

INTERMEDIATE WATERS

BOTTOM WATERS

# *36. The life of a berg*

**WHEN ICE FROM GLACIERS** and ice sheets eventually reaches the coastline, it breaks off to form icebergs. These icebergs vary hugely in size and shape. The largest are the great tabular bergs that calve from the ice shelves. These are flat-topped and can be vast – sometimes the size of a small country. Bergs from glaciers tend to be smaller and more pointed and can form fantastic shapes with pinnacles; they can resemble castles. The seawater around the coast of Antarctica is almost always extremely cold – usually -1 to -2°C – and is kept from freezing by its saltiness. So the bergs in the southern seas do not melt rapidly and can remain in the coastal waters for many years. They follow the Antarctic Coastal Current, a strong current that flows anticlockwise, in the opposite direction to the main Antarctic Circumpolar Current (ACC) further north. Locked into this flow, bergs often take years to circumnavigate Antarctica before they reach the Weddell Sea, where the rocky arm of the Antarctic Peninsula bars their westward journey. Here they are spun out northwards into the Scotia Sea and the South Atlantic, in 'iceberg alley'. In these warmer waters, the great bergs eventually melt and die, adding fresh water and nutrients into the ocean. The largest tabular bergs will take longest to melt and often travel north of South Georgia and the Falkland Islands, causing a risk to the shipping lanes of the oceans. These big bergs are tracked by satellite, tens or hundreds each year, and the map here shows those paths as the bergs make their way around the continent and out into the warmer waters. The colour of each line is based upon the time and year the berg was tracked, so each berg shows as a separate colour.

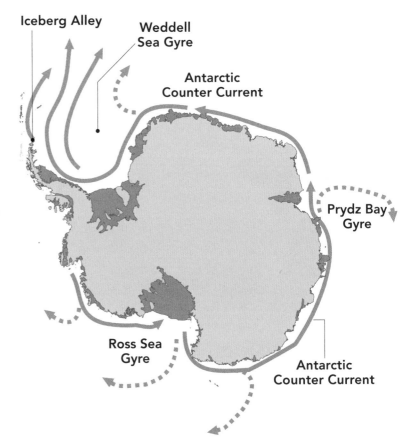

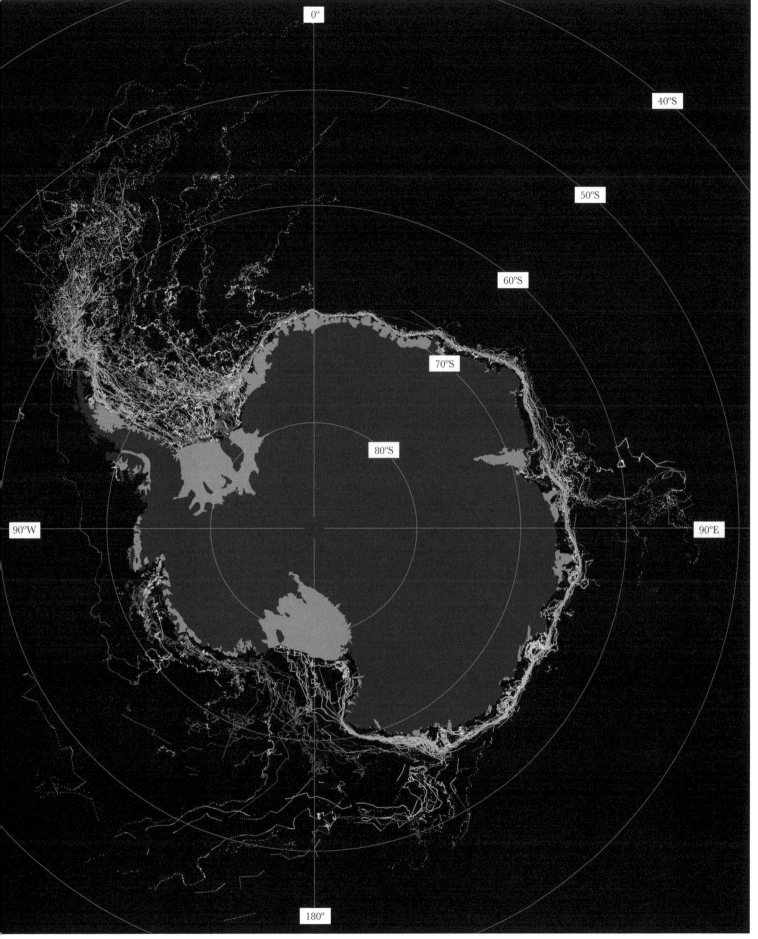

# 37. The green ocean

**IN THE ANTARCTIC SUMMER**, the Southern Ocean is one of the most productive bodies of water on Earth. The spatial pattern of that productivity is very variable, as shown on this map. 'Productivity' is a term used to specify how much photosynthesis there is in the ocean. On land, photosynthesis is the process whereby plants take water, sunlight and nutrients and turn them into energy. In the oceans, the process is the same but, instead of plants, the organisms that photosynthesize are phytoplankton – microscopic floating algae too small to be seen by the naked eye but sometimes present in high enough concentrations to turn the sea green. This rich soup of organic matter is the basis of the food chain and supports all life in the ocean.

Here, in the cool Southern Ocean, the combination of strong winds, the upwelling of deep water and the mixing of different water masses brings nutrients to the surface which are consumed by the phytoplankton. As these factors are geographically localized, the pattern of ocean productivity is unevenly spread. Major geographic influences include the mixing of cooler and warmer waters around the polar front, glacial meltwaters and the run-off from islands. All these shape the pattern of productivity, with hotspots around the Scotia Sea, the Antarctic Peninsula and Sub-Antarctic

islands. Notice the contrast between the cooler waters near Antarctica and the warmer waters near the equator, which are often almost barren by comparison.

One of the other fascinating things about this productivity is the individual shapes of the organisms. Phytoplankton come in a bewildering range of forms and types. The circle of illustrations below depicts some of the major types and groups.

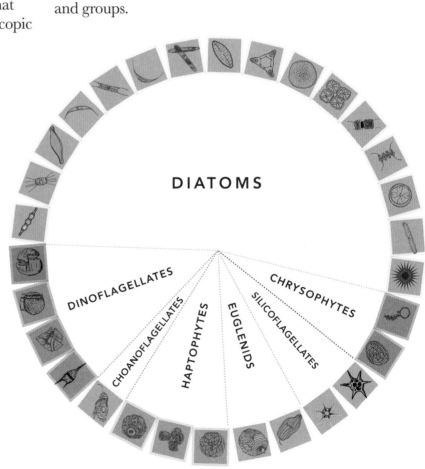

DIATOMS

DINOFLAGELLATES

CHOANOFLAGELLATES

HAPTOPHYTES

EUGLENIDS

SILICOFLAGELLATES

CHRYSOPHYTES

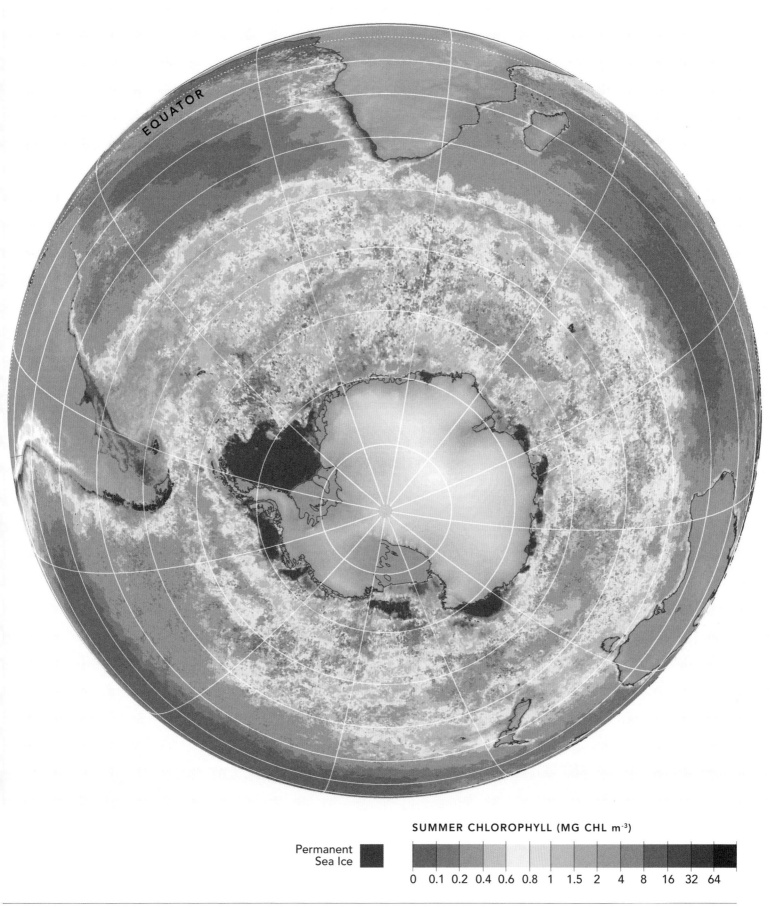

EQUATOR

**SUMMER CHLOROPHYLL (MG CHL m⁻³)**

Permanent
Sea Ice

0   0.1   0.2   0.4   0.6   0.8   1   1.5   2   4   8   16   32   64

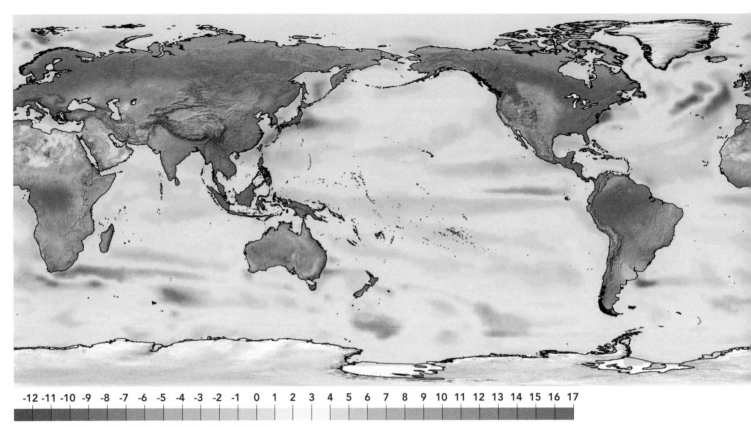

-12 -11 -10 -9 -8 -7 -6 -5 -4 -3 -2 -1 0 1 2 3 4 5 6 7 8 9 10 11 12 13 14 15 16 17

CUMULATIVE OCEAN HEAT UPTAKE ($10^9 J/m^2$)

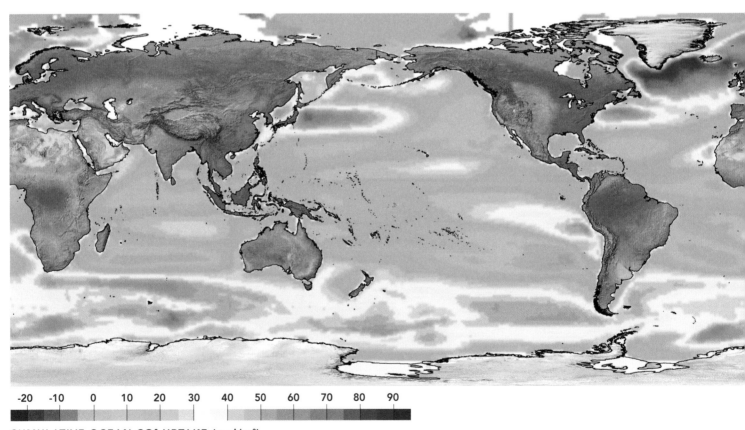

-20 -10 0 10 20 30 40 50 60 70 80 90

CUMULATIVE OCEAN $CO_2$ UPTAKE (mol/m²)

# 38. Earth's lungs

**THE SOUTHERN OCEAN** acts like a giant air-conditioning system, moderating the temperature of the planet.

Since humanity started burning oil, coal and other fossil fuels, we have been putting carbon dioxide into our atmosphere. This carbon traps heat in the atmosphere and produces a phenomenon called the 'greenhouse effect', which in turn leads to global warming. Luckily for us, several processes on the planet take away some of the extra carbon and heat, mitigating the worst effects of global warming. Places where this happens are commonly called 'sinks', and by far the greatest of these sinks is the Southern Ocean. Its rough, turbulent character helps to dissolve the carbon and transport it into the depths.

Since the beginning of the Industrial Revolution, it is estimated that the Southern Ocean has taken up 43 per cent of the extra carbon put into the atmosphere by humans. It has also absorbed 75 per cent of the extra heat caused by the greenhouse effect. Scientists estimate that global temperatures have risen by 0.8°C in little over a century, but if it were not for the Southern Ocean, the increase would be four times higher.

The maps show the carbon and heat uptakes of the world's oceans between 1870 and 1995. Notice the dominance of the Southern Ocean in both these plots. Scientists recently showed that, paradoxically, the higher winds in the Southern Ocean have started to reduce the uptake of carbon in the region. This unexpected discovery is a worrying trend which might reduce the dampening effect of the ocean and speed up the rate of climate change around the globe.

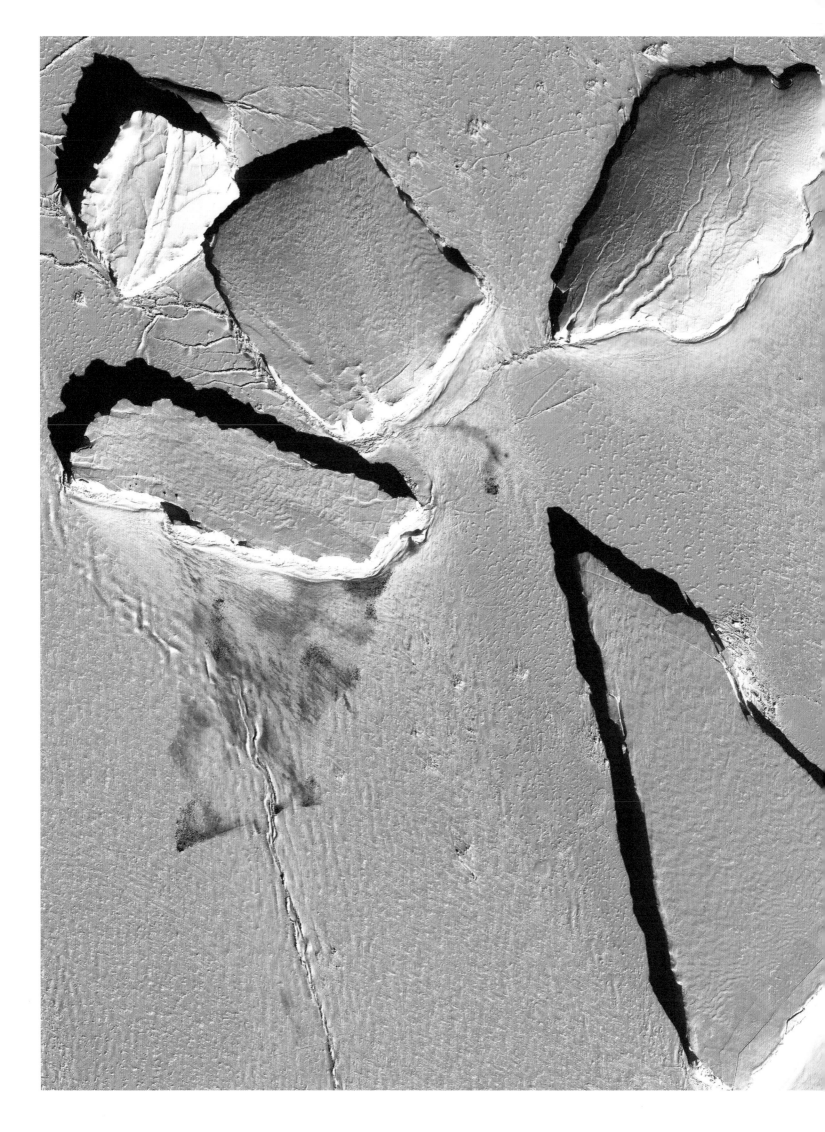

# CHAPTER 6

# *Wildlife*

39. Keystone krill  *108*

40. The realm of the emperor  *110*

41. An ocean of penguins  *112*

42. International seal travels  *114*

43. The blood-red sea  *116*

44. The great wanderers  *118*

45. The richest place on earth  *120*

# 39. *Keystone krill*

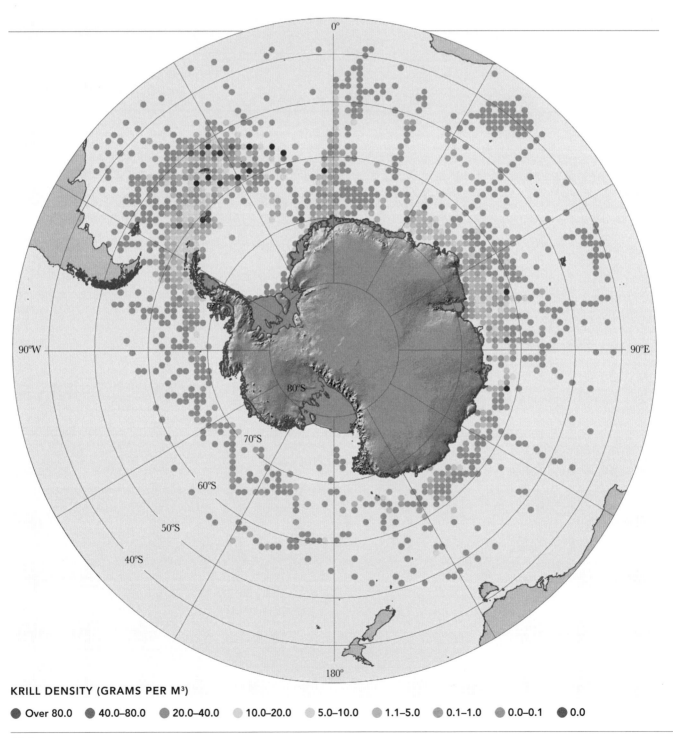

**KRILL DENSITY (GRAMS PER M³)**

● Over 80.0    ● 40.0–80.0    ● 20.0–40.0    ● 10.0–20.0    ● 5.0–10.0    ● 1.1–5.0    ● 0.1–1.0    ● 0.0–0.1    ● 0.0

**KRILL IS THE MOST** important creature in the Southern Ocean. This small, shrimp-like crustacean is critical to marine ecosystems. Krill eat the tiny phytoplankton at the bottom of the food chain, and virtually everything else eats krill – from fish to penguins, from seals to the great whales.

Lucky, then, that there are lots of krill. In the Southern Ocean, where this little creature abounds in huge numbers, it's estimated that there are over 350 million tonnes of them. This is probably the greatest biomass of any single living creature on the planet. You wouldn't want to be one, though; half of all krill are thought to be eaten every year by the multitude of predators that feast on this bounteous crustacean.

Krill tend to live in the deeper layers of the ocean during the day (up to 500 metres deep).

Each night, they rise to the surface to feed on the abundant phytoplankton there. They make this daily journey to avoid predators, which find it easier to catch them in the light. Before dawn, they sink back into the safety of the depths. This vertical movement is said to be the largest daily migration of biomass on Earth.

The map shows the distribution of krill in the Southern Ocean, based on catch records. It's not easy to work out. Krill live in swarms, and those swarms move, so when you cast your net over the side to try to work out how much krill there is, you never get the same density in the same place twice, so you have to take a lot of samples to work out the true abundance in any area. Each dot on the map represents the average of all the catches in a 100-square-kilometre cell. The redder the cell, the more krill there are in that area. The greatest density is around the Scotia Sea region, and it is here that the densest congregations of predators on Earth are also found (see The Richest Place on Earth map, pages 120–21).

For us, it is important to understand and map where the krill are, as humans also prey on them. Fishing fleets catch more and more each year. At present, we take around 200,000 tonnes annually, mostly for animal feed, but they are also used in an important and expanding industry in pharmaceuticals and cosmetics. Refined krill oil is used in medicines for heart disease and high cholesterol, as a health-food supplement and in anti-ageing serum. The total catch is not a large amount, considering that the overall biomass of the species is so high. The Antarctic fishing industry is one of the best regulated in the world, and it tries very hard to ensure that fishing in the Southern Ocean does not impact upon its wildlife. But there is concern that, in future, if the fishing fleets catch large amounts of krill locally, it may mean that penguin and seal colonies near those fishing grounds will not have enough food to raise their young.

KILLER WHALE
LEOPARD SEAL
ELEPHANT SEAL
PENGUINS
SEABIRDS
SEALS
WHALES
FISH
KRILL
HUMANS
SQUID
ZOO PLANKTON
PHYTOPLANKTON

# 40. *The realm of the emperor*

**EMPEROR PENGUINS ARE** in trouble. This is a bird that depends upon sea ice to breed, and the first thing to disappear as the world warms will be the sea ice.

It's already happened in the Arctic, where rising temperatures have led to a rapid decline in the ice, with lower ice-cover over winter and summers becoming almost ice-free. Scientists predict that over the next century, as climate change pushes further south into the Antarctic, the sea ice here will also be lost.

This will have catastrophic consequences for emperor penguins. This iconic species uses the frozen sea as a platform to breed on, and for much of the year emperor penguins forage for food beneath the ice. In short, no sea ice means no emperors.

Over the last decade, scientists have tried to model how much and how quickly the emperor penguin population will decline, but it's difficult. These birds breed in such extreme environments that our knowledge of how they react to change is sketchy. Only about half of the fifty or so colonies that exist like a necklace around the Antarctic coastline have ever been visited by humans – the rest have

been found and studied only through satellite imagery. Only a handful of the others are visited on a regular basis, and those tend to be the ones that exist on the most stable ice, in the areas that are easiest to access from research stations. Add this to the fact that our models of future climate change are imperfect, and it is probably best to say that the estimates of where and how the population will decline are a work in progress.

This map shows our best estimate of what will happen to emperor penguin populations by the end of this century. It suggests that, by 2100, the population of emperor penguins will be only 30 per cent of today's numbers, with thirty-four of the existing fifty-three colonies extinct, thirteen declining rapidly and only six stable or growing. All of these six sites are in the deep south, around the Ross Sea. What will happen after 2100 has not yet been assessed.

But there is little doubt that emperors will struggle with their empire melting beneath their feet.

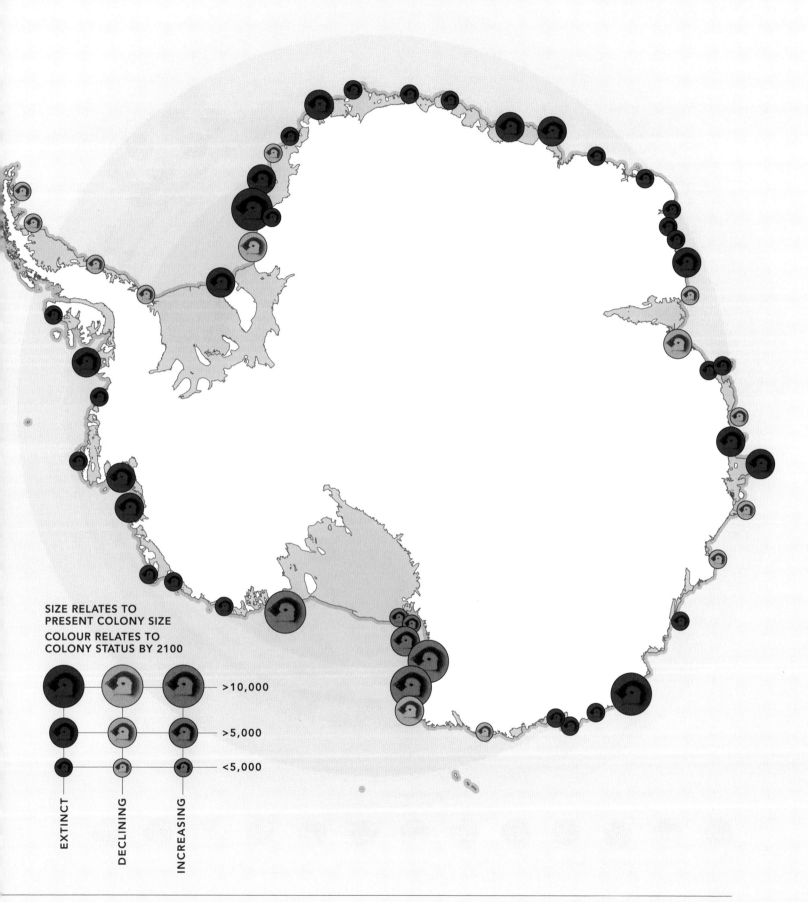

SIZE RELATES TO
PRESENT COLONY SIZE

COLOUR RELATES TO
COLONY STATUS BY 2100

>10,000

>5,000

<5,000

EXTINCT

DECLINING

INCREASING

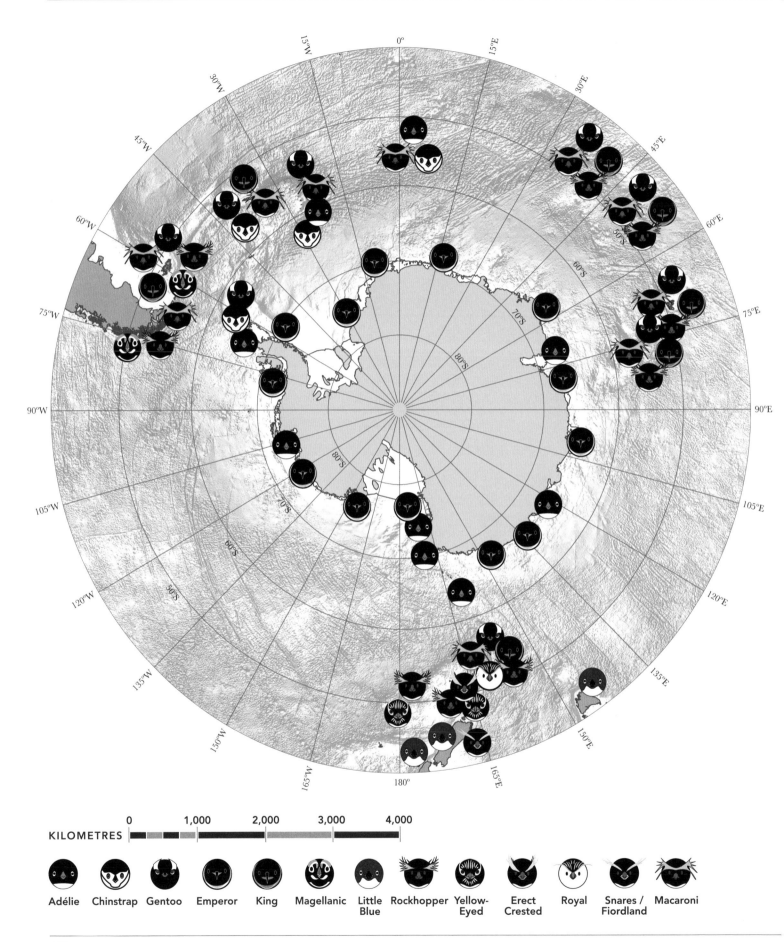

KILOMETRES

0    1,000    2,000    3,000    4,000

Adélie  Chinstrap  Gentoo  Emperor  King  Magellanic  Little Blue  Rockhopper  Yellow-Eyed  Erect Crested  Royal  Snares / Fiordland  Macaroni

# 41. An ocean of penguins

**PENGUINS DOMINATE THE ECOSYSTEM** of the Southern Ocean. Every island or storm-lashed speck of rock poking out of the sea has penguin colonies. The number of penguins on these islands is huge. Often colonies are hundreds of thousands strong; sometimes, over a million birds can reside in a single colony. Packed on to tiny amounts of land, often on precipitous slopes of the steep islands, each colony is a raucous, flapping, smelly cacophony – a celebration of the rich, bounteous life of the Southern Ocean.

Of the seventeen species of penguin, only four live and breed in Antarctica: the emperor, Adélie, Gentoo and Chinstrap. Only the emperor and Adélie brave the harsh, ice-bound coasts of continental Antarctica south of the Antarctic Peninsula. Another eleven species live on the islands in and around the Southern Ocean, mostly around the productive and turbulent polar front. The other three types breed further to the north, around the coasts of South America and Africa, and as far north as the Galapagos Islands. The area just south of New Zealand has the greatest diversity of species. Ten different types of penguin breed on South Island or the islands to the south of it. Several of these types, such as the royal, erect crested and Fiordland penguins, live nowhere else.

**PENGUIN PARADE**

HEIGHT (CM)

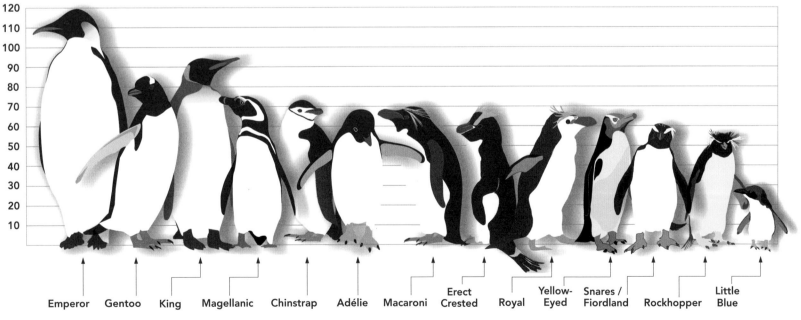

Emperor   Gentoo   King   Magellanic   Chinstrap   Adélie   Macaroni   Erect Crested   Royal   Yellow-Eyed   Snares / Fiordland   Rockhopper   Little Blue

# 42. *International seal travels*

**UNDERSTANDING THE LIFE** of seals in the Southern Ocean is challenging. Four of the five seal species that live and breed here spend much of their lives in the sea ice, which makes them extremely difficult to study. For scientists to move over or through this ever-changing icy mosaic of frozen icebergs and pack ice is gruelling, expensive and downright dangerous. The ecosystem the sea ice supports is rich, with many penguins, seabirds, seals, whales and fish dependent upon it. So getting a glimpse into how it works is a crucial objective for scientists. But how do you study something if you cannot get to it? Enter the elephant seal. The southern elephant seal is the world's largest – adult males can grow up to 5.5 metres long and weigh up to 5 tonnes. It breeds on the remote islands of the Southern Ocean, but feeds throughout the southern seas, wherever there is food. Adults can often be found in the pack ice, as much at home here as in the warmer waters further north. The elephant seal travels immensely long distances for food. If you think that you have a long commute, pause a moment to consider the elephant seal, who may swim over 10,000 kilometres in search of food. The longest recorded track in the database is a seal that swam out from a breeding colony on South Georgia and travelled 17,600 kilometres.

Researchers use elephant seals as a surrogate science platform; by attaching tracking devices to them they can discover not just where the seals go to forage for food, but also measure temperature, salinity and the chemical composition of the water they travel through. Elephant seals can dive to incredible depths in search of food, so by fitting depth sensors, we can even use them to map the topography of the sea floor in places ships cannot reach. Luckily, the seals are so large that all this instrumentation is an insignificant burden to them.

Over the past few years, there has been a great international effort to use seals in this way, and many nations active in Antarctic science have been tagging them. The map here shows the results of that work. Each thin, coloured line represents an individual seal's journey. Lines are coloured according to the nationality of the science team that placed the tag on the seal, evidence of how multinational the scientific community is. More recently, Weddell seals, which breed further to the south, deep within the pack, have also been tagged; their tracks are shown as dashed lines. Some of the main breeding areas where the instruments are attached are indicated by seal symbols.

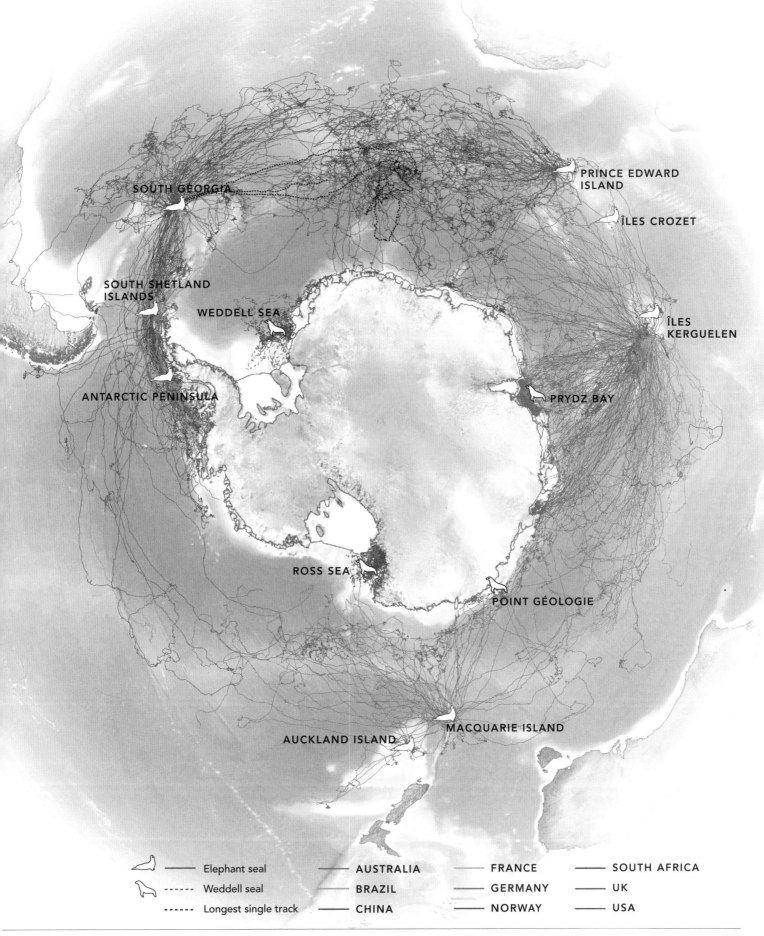

SOUTH GEORGIA

PRINCE EDWARD
ISLAND

ÎLES CROZET

SOUTH SHETLAND
ISLANDS

WEDDELL SEA

ÎLES
KERGUELEN

ANTARCTIC PENINSULA

PRYDZ BAY

ROSS SEA

POINT GÉOLOGIE

AUCKLAND ISLAND

MACQUARIE ISLAND

| | | | |
|---|---|---|---|
| Elephant seal | AUSTRALIA | FRANCE | SOUTH AFRICA |
| Weddell seal | BRAZIL | GERMANY | UK |
| Longest single track | CHINA | NORWAY | USA |

# 43. The blood-red sea

**THE SLAUGHTER OF** the great whales in the Southern Ocean must be one of the most sobering stories in the history of humanity's interaction with our natural environment. These intelligent creatures, which include the largest animals ever to exist on the planet, were once common in our seas and oceans. Hundreds of thousands of them swam in the Southern Ocean alone, but over the course of the last two centuries they have been systematically hunted to within a hair's breadth of extinction.

Between 1915 and 1984, around 1.6 million whales were caught in the Southern Ocean, with a combined biomass equivalent to that of almost the whole human race.

It started with the easiest whale to catch: the right whale. It was slow, fat and swam on the surface of the sea around coastal waters, so for many years it was the only whale that could be caught. Its name literally means the 'right' whale to hunt. As ships became faster, other slower but more powerful

beasts such as gray whales and sperm whales became targets. By the late 1800s, these whales had already been almost fished out of the more northern waters, and the numbers of right whales had decreased so much they were all but extinct. From 1915, due to international pressure from many whaling nations who were beginning to realize that the rate of whaling was unsustainable, all whale catches have been recorded. The map opposite is based on these records. Notice that there is no right whale map, as by this time virtually no right whales existed.

The hunting has now almost stopped. In 1986, the International Whaling Commission, which includes the majority of nations involved in whaling, banned hunting. It is technically only suspended and, as some nations never joined or never agreed to the commission ruling, some fleets still hunt these great beasts. Whether whaling will ever again resume on the scale of the early twentieth century is a thorny issue for

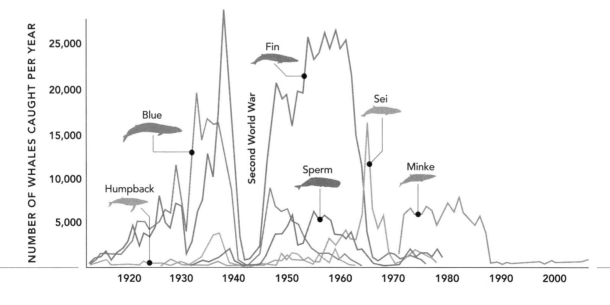

the international community, for, although many western nations regard the killing of whales as a needless slaughter, some cultures still see whales as a natural resource to be harvested.

The maps show the distribution of whale catches in the Southern Ocean; the darker the red, the more whales have been caught. The large map in the centre of the circle shows the combined catch, while the smaller maps show the pattern by species. Notice that each whale type prefers different waters: sei are subtropical, minke whales love colder waters, while blues and fins congregate around the frontal zones. By the time these data were collected, many of the humpback and sperm whales had already been taken, so there are fewer catches recorded for them on these maps.

NUMBER OF WHALES CAUGHT
PER 1x1 DEGREE OF OCEAN

1 2 4 8 16 32 64 100 128 256 512

SEI WHALES

BLUE WHALES

FIN WHALES

ALL WHALES

SPERM WHALES

HUMPBACK WHALES

MINKE WHALES

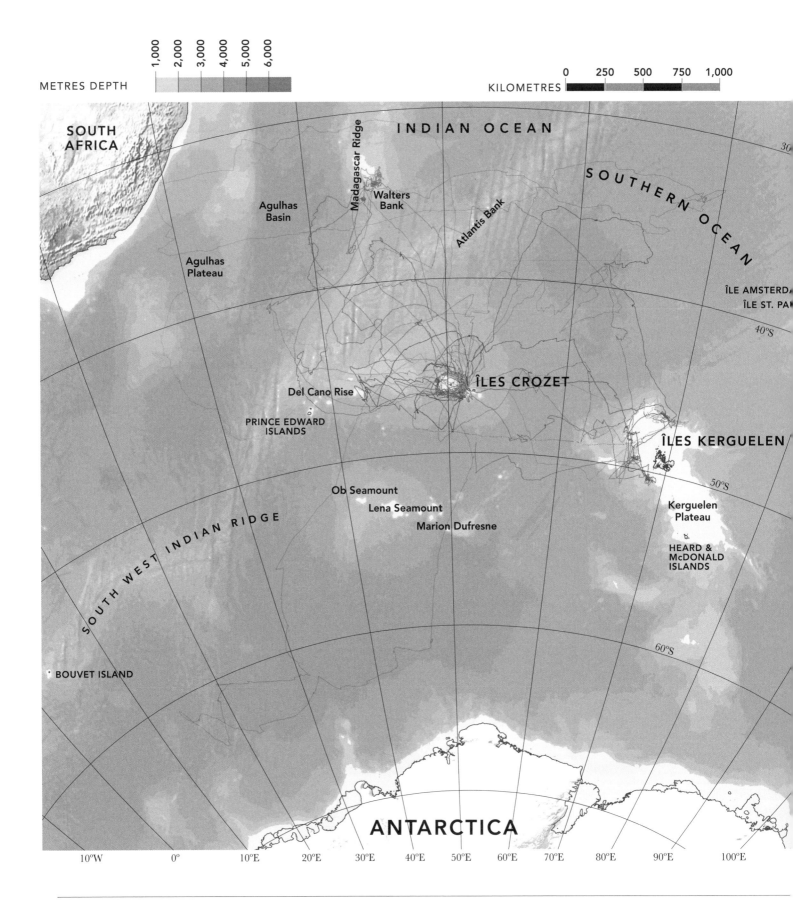

METRES DEPTH

1,000 2,000 3,000 4,000 5,000 6,000

KILOMETRES

0 250 500 750 1,000

SOUTH
AFRICA

INDIAN OCEAN

SOUTHERN OCEAN

Madagascar Ridge

Walters
Bank

Agulhas
Basin

Atlantis Bank

Agulhas
Plateau

ÎLE AMSTERD
ÎLE ST. PA

30

40°S

Del Cano Rise

ÎLES CROZET

PRINCE EDWARD
ISLANDS

ÎLES KERGUELEN

50°S

Ob Seamount

Lena Seamount

Marion Dufresne

Kerguelen
Plateau

HEARD &
McDONALD
ISLANDS

SOUTH WEST INDIAN RIDGE

60°S

BOUVET ISLAND

ANTARCTICA

10°W    0°    10°E    20°E    30°E    40°E    50°E    60°E    70°E    80°E    90°E    100°E

# 44. *The great wanderers*

**WITH THE LARGEST WINGSPAN** of any bird, the wandering albatross is truly an awesome creature. Weighing up to 12 kilograms, with a wingspan of 3.5 metres, the 'Wanderer' is clumsy on land and sometimes has difficulty getting airborne if there is not a strong wind blowing, but once in the air it is the most elegant of flyers. It glides on the strong winds of the Southern Ocean with an effortless grace, rarely flapping those huge wings and using the barest amount of energy to stay aloft. The species can travel huge distances, averaging almost 1,000 kilometres a day and sometimes 10,000 kilometres in a single foraging trip. Some birds have been recorded to have circumnavigated the entire Southern Ocean, almost 30,000 kilometres, in just forty-six days.

Where they go and why they go there is still something of a mystery but, with recent technological advances in tracking devices, we are beginning to understand the rovings of this enigmatic bird. This map shows the tracks of twenty-eight birds that were tagged by French researchers in the Crozet Islands in the Indian Ocean sector of the Southern Ocean. This small, remote and rocky group of islands is a haven for wildlife, providing a breeding location for thousands of penguins, seals and flying seabirds in the middle of the tumultuous Southern Ocean.

Over fifty thousand individual GPS records are plotted on this map, each bird in a different colour. Judging by these lines, the species lives up to its name, wandering seemingly at random to all points of the compass. On closer inspection, the path of many of the tracks concentrates on physical features such as the edges of the submarine plateaus around the Crozet and Kerguelen groups of islands and around seamounts (undersea mountains). These areas are likely to have high levels of nutrients and productivity brought up from deeper waters, and the albatrosses are actively seeking out these rich waters in order to feed.

# 45. The richest place on earth

**THEY SAY THAT** the coastal strip of South Georgia has a greater density of wildlife than anywhere else on the planet. A staggering five million fur seals, millions of penguins, tens of thousands of giant elephant seals and some of the world's largest albatross and flying seabird colonies can be found here. You cannot step on to the beaches of South Georgia without being staggered by the sheer mass of life. These animals breed here, drawn by the rich, cold, productive waters that circle the island, waters that are fertilized by deep upwelling currents and outflowing sediment from the 160 or so glaciers that flow off the steep, mountainous interior into the sea.

The island and its wildlife have had a turbulent history. Captain James Cook was the first to make landfall here, in 1775, and soon afterwards the sealers came, hunting the multitude of fur and elephant seals for their blubber and skins. Over the next century and a half the sealers all but annihilated the population of this animal. By the mid-twentieth century, the number of fur seals on South Georgia had dropped from several million to fewer than fifty individuals. When the seal population had diminished, the whalers came. Arriving in the early part of the twentieth century, they established six industrial whaling stations, making the island the hub of the global whaling industry. It was here that fleets of whale boats set out to catch the great sea creatures, and it was back to South Georgia that the dead whales were brought for processing. Hundreds of thousands were cut up and boiled down for oil in the whaling stations, until the whales, too, were on the

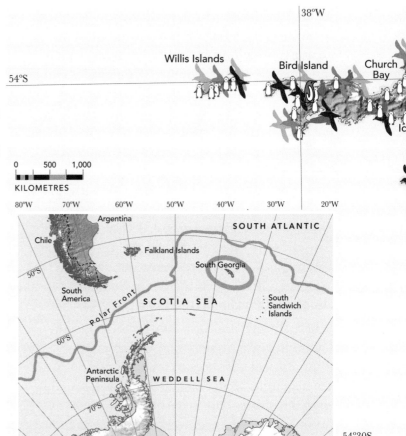

brink of extinction. And it wasn't only the seals and whales that suffered. The whalers brought reindeer for food, and rats and mice inadvertently escaped from their ships onto land. These alien species had a devastating effect on the globally important seabird colonies that covered the island. The trampling of reindeer and the predation of the eggs and chicks by rats made much of the mainland a no-go area for the many species of albatross and petrel that had been accustomed to breed here.

**120**

The whaling stations have been abandoned now; they were closed after the moratorium on Southern Ocean whaling in 1965. The animals the whalers brought with them – the reindeer, the mice and the rats – have been eradicated in an attempt to allow native birds to recover. The seals have returned, multiplying rapidly to fill the beaches, and the populations of great whales that surround the island are recovering. Hopefully, the albatross will recover, too, but this may be a false dawn. Climate change is rapidly altering the icy heart of the island and, as the temperature rises, the glaciers retreat. South Georgia is one of the fastest-deglaciating places on Earth: some glaciers have retreated up to 14 kilometres; others have disappeared altogether. This could have critical consequences for the wildlife there; the glaciers erode the rocks and deposit the silt into the ocean, turning the surrounding seas a milky-blue colour. These sediments provide around half of the nutrients that feed this rich marine ecosystem, but when the glaciers disappear, as they will in the coming decades, that source of nutrients will die. What will happen then to the abundance of wildlife that makes South Georgia unique and globally important is, as yet, unknown.

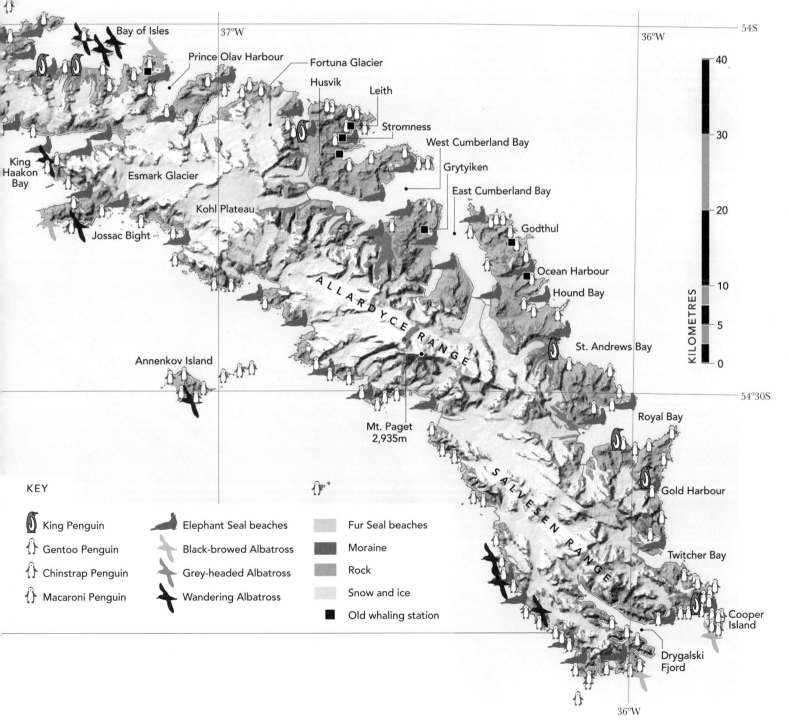

KEY

King Penguin
Gentoo Penguin
Chinstrap Penguin
Macaroni Penguin

Elephant Seal beaches
Black-browed Albatross
Grey-headed Albatross
Wandering Albatross

Fur Seal beaches
Moraine
Rock
Snow and ice
Old whaling station

# CHAPTER 7

# *People*

46. Going south *124*

47. Who lives there? *126*

48. Sweet home Antarctica *128*

49. Pieces of pie *132*

50. Who owns it? *134*

51. Mac Town *136*

52. Mobile home *138*

53. International antics *142*

54. Antarctic skies *144*

55. Exploiting the ocean *146*

56. Tourist hub *148*

# 46. Going south

**GETTING TO ANTARCTICA** today is easier than it used to be. Now, most people fly directly to the continent. Although there are no major airports that we would recognize the names of, there are a number of gravel (black squares on the map) or hard-ice (black dots) runways that accept long-distance jets or turbo-prop planes. Ships are still important, as they carry the majority of fuel, supplies and equipment into the continent.

Each of the southern continents – Africa, Australia and South America – has specific 'gateway' cities that act as logistic hubs, transporting people and cargo south. Each city supports the research stations directly south of it. Cape Town in South Africa supplies the stations of Dronning Maud Land; Hobart, on the Australian island of Tasmania, supplies most of the rest of East Antarctica; Canterbury in New Zealand is the gateway to the Ross Sea region; while the Chilean and Argentinian cities of Punta Arenas and Ushuaia support the bases on the Antarctic Peninsula.

Virtually all the aircraft and ships that visit Antarctica, either to resupply the research stations or to bring tourists to the continent, set out from one of these five gateway cities. The only other point of departure is Port Stanley in the Falkland Islands, which still supplies some of the UK's research stations.

The planes that bring people to Antarctica need to cross the Southern Ocean, which is almost

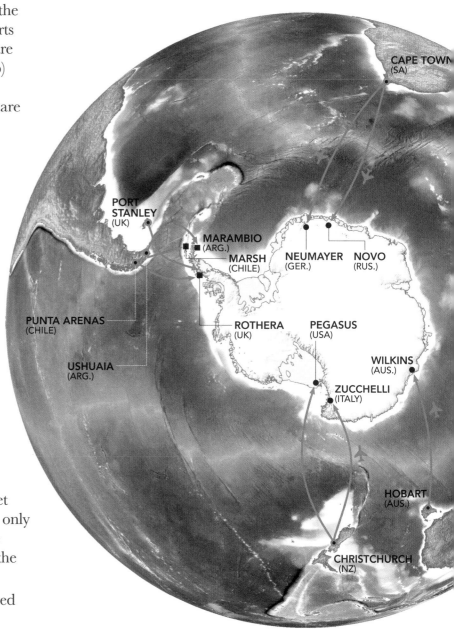

CAPE TOWN
(SA)

PORT
STANLEY
(UK)

MARAMBIO
(ARG.)

MARSH
(CHILE)

NEUMAYER
(GER.)

NOVO
(RUS.)

PUNTA ARENAS
(CHILE)

ROTHERA
(UK)

PEGASUS
(USA)

WILKINS
(AUS.)

USHUAIA
(ARG.)

ZUCCHELLI
(ITALY)

HOBART
(AUS.)

CHRISTCHURCH
(NZ)

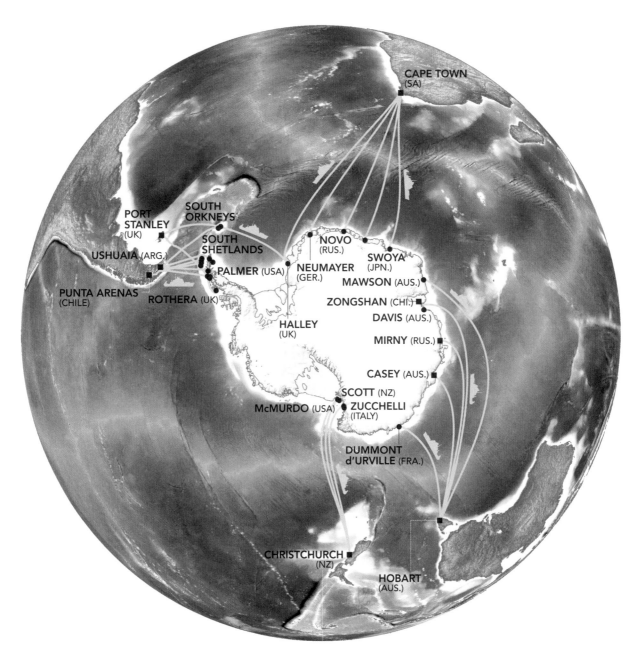

1,000 kilometres across at its narrowest stretch between South America and the Peninsula. To fly from Cape Town or Canterbury is further, so only large, intercontinental aircraft, usually ex-military types such as Lockheed C-130s or massive Russian Ilyushin IL-76 cargo planes, make that journey.

Tourist ships come in all shapes and sizes, but mostly ply their trade in the relatively safer seas surrounding the Antarctic Peninsula, where there is open water in summer. Research ships often tend to be better equipped and have stronger, ice-strengthened hulls as a protection against the ice. They frequently travel further south, penetrating into heavy pack ice to reach the research stations.

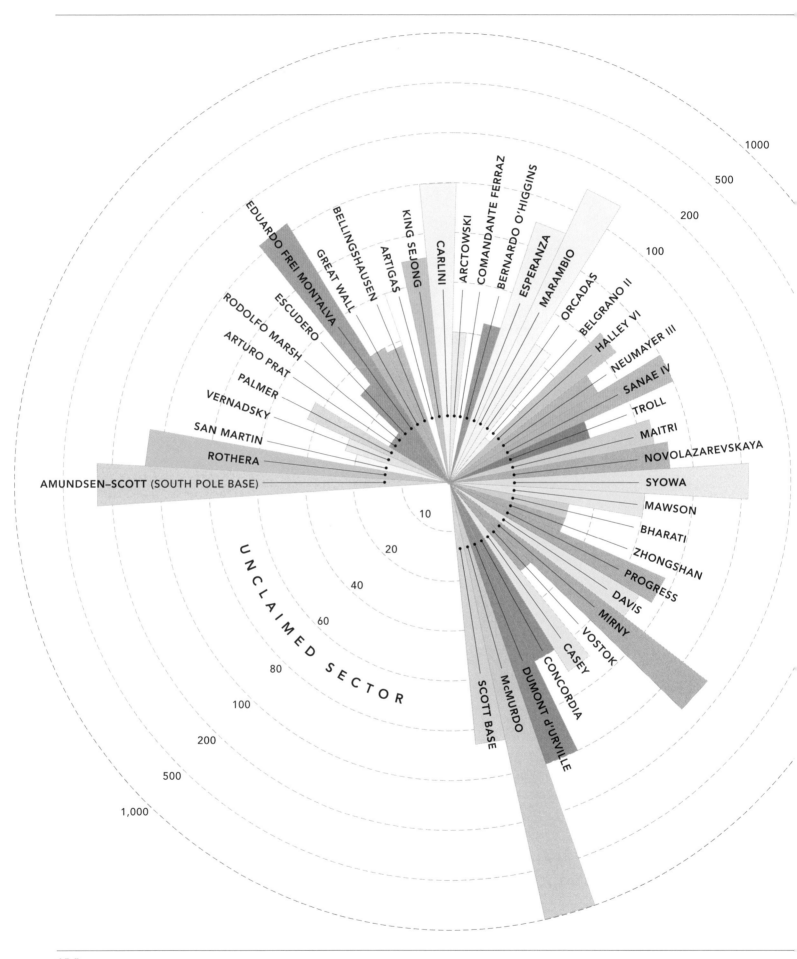

1000
500
200
100

EDUARDO FREI MONTALVA
BELLINGSHAUSEN
GREAT WALL
ARTIGAS
KING SEJONG
CARLINI
ARCTOWSKI
COMANDANTE FERRAZ
BERNARDO O'HIGGINS
ESPERANZA
MARAMBIO
ORCADAS
BELGRANO II
HALLEY VI
NEUMAYER III
SANAE IV
TROLL
MAITRI
NOVOLAZAREVSKAYA
SYOWA
MAWSON
BHARATI
ZHONGSHAN
PROGRESS
DAVIS
MIRNY
VOSTOK
CASEY
CONCORDIA
DUMONT d'URVILLE
McMURDO
SCOTT BASE

ESCUDERO
RODOLFO MARSH
ARTURO PRAT
PALMER
VERNADSKY
SAN MARTIN
ROTHERA
AMUNDSEN–SCOTT (SOUTH POLE BASE)

U N C L A I M E D   S E C T O R

10
20
40
60
80
100
200
500
1,000

# 47. Who lives there?

**WINTER TOTAL: 1,100**

500 250 100 0

**SUMMER TOTAL: 3,368**
(4,100 INCLUDING NON-WINTERING BASES)

0 100 250 500 1,000 1,500

WINTER POPULATION

SUMMER POPULATION

USA
RUSSIA
ARGENTINA
CHILE
AUSTRALIA
UK
FRANCE/ITALY
JAPAN
INDIA
NEW ZEALAND
SOUTH AFRICA
CHINA
SOUTH KOREA
URUGUAY
GERMANY
BRAZIL
POLAND
NORWAY
UKRAINE

**ANTARCTICA HAS NO** permanent population and is the only continent on Earth that has no indigenous people. Up until around seventy years ago, it was totally uninhabited. The first military and research stations were set up in the last years of the Second World War and proliferated in the 1950s and 1960s, mainly established by the superpowers of the day and the nations who, at that time, claimed sovereignty over parts of the continent. Many of the original bases were run by the military, but after the signing of the Antarctic Treaty in 1959, scientists and their support staff became the main residents of those stations. As more countries have taken an interest in Antarctica and more nations have signed the Antarctic Treaty, more research stations from many more countries have been built. Today, twenty-nine nations have stations in Antarctica. In the summer, the population of the continent can be almost four thousand people, while in the winter the population is barely a thousand. The graphic gives an indication of how many people from each country live here in winter and in summer. By far the largest number are from the US – over twice as many as from any other nation in both summer and winter. The second largest contingent is from Russia, closely followed by those from the South American rivals of Argentina and Chile. Both these nations have territorial claims, as do the next largest populations, from Australia and France.

The larger figure on the graphic gives an indication of each of the main research stations' maximum capacity. The figure is roughly orientated with the position of each station around Antarctica. The gap between Amundsen–Scott (the official name of the South Pole station) and Scott Base relates to the uninhabited zone between the Ross Sea and the Antarctic Peninsula where there are no settlements of any kind.

# 48. Sweet home Antarctica

**ACCOMMODATION IN ANTARCTICA** has evolved significantly since the wooden huts of the early explorers. The space-age designs of recently built research stations are unique and impressive.

The harsh conditions of the polar environment mean that the modern station faces several challenges. Cold is the obvious one, and all buildings need to be extremely well insulated, with thick cladding an essential part of any polar base. But a more serious problem is snow accumulation. The snow in Antarctica never melts, so buildings on the surface are soon covered and buried, eventually to be crushed beneath thousands of tonnes of snow and ice. To get around this, most modern stations are now built on stilts and some are jacked up every year to ensure that the snow blows underneath rather than up against them. Many of them also have rounded edges and are orientated into the prevailing wind to try to reduce wind noise and the vibration of the base in the howling gales of the Antarctic winter.

One of the most serious threats to people living on the continent is a fire on base. In Antarctica, if a station burns down, the nearest habitation to that base is often hundreds of kilometres away, which is a serious problem in the inhospitable conditions surrounding the buildings. There needs to be somewhere close by to act as a shelter for the station occupants to survive until help arrives. Because of this risk, many stations are modular in design; that way, if one part of the base becomes uninhabitable, personnel can simply move to a different building that a fire cannot reach. This is especially true of the overwintering stations in the deep south, such as Halley VI, Concordia and Amundsen–Scott (South Pole) station.

## Grand Designs Antarctica

### HALLEY VI *(Great Britain)*

Halley VI is located on a floating ice shelf, which is continually moving towards the coast. So Halley has been designed to be broken apart and towed inland. The station consists of eight interconnected modules attached together in a line. The research here is mainly atmospheric science, and it was here, in 1984, that the hole in the ozone layer was discovered (see pages 74–77).

### CONCORDIA *(France/Italy)*

The French/Italian station of Concordia is built on one of the highest and most isolated parts of the Antarctic ice sheet. At an altitude of 3.2 kilometres above sea level and over 1,000 kilometres from the coast the location is extremely cold ,with an average temperature of -54.5°C. In winter the thermometer sometimes plummets to below -80°C and in the summer it rarely rises above -25°C. The base consists of two drum-shaped, three-storey buildings, and a smaller technical facility connected by enclosed walkways. Like several other stations situated on the ice, Concordia stands on legs, raising it above the surface and allowing the snow to blow beneath it.

*(Continued overleaf)*

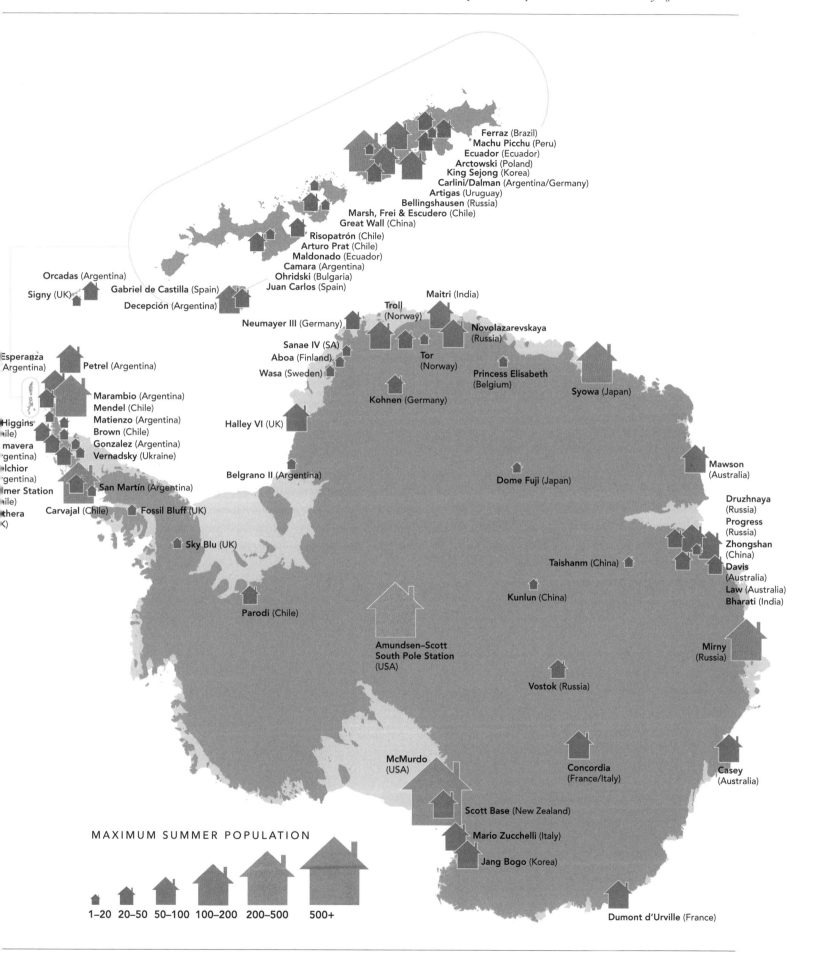

Ferraz (Brazil)
Machu Picchu (Peru)
Ecuador (Ecuador)
Arctowski (Poland)
King Sejong (Korea)
Carlini/Dalman (Argentina/Germany)
Artigas (Uruguay)
Bellingshausen (Russia)
Marsh, Frei & Escudero (Chile)
Great Wall (China)
Risopatrón (Chile)
Arturo Prat (Chile)
Maldonado (Ecuador)
Camara (Argentina)
Ohridski (Bulgaria)
Juan Carlos (Spain)

Orcadas (Argentina)
Signy (UK)
Gabriel de Castilla (Spain)
Decepción (Argentina)
Neumayer III (Germany)
Sanae IV (SA)
Aboa (Finland)
Wasa (Sweden)

Maitri (India)
Troll (Norway)
Novolazarevskaya (Russia)
Tor (Norway)
Princess Elisabeth (Belgium)
Syowa (Japan)
Kohnen (Germany)

Esperanza (Argentina)
Petrel (Argentina)
Marambio (Argentina)
Mendel (Chile)
Matienzo (Argentina)
Brown (Chile)
Gonzalez (Argentina)
Vernadsky (Ukraine)
Higgins (Chile)
Primavera (Argentina)
Melchior (Argentina)
Palmer Station (Chile)
Rothera (UK)
Carvajal (Chile)
San Martín (Argentina)
Fossil Bluff (UK)
Belgrano II (Argentina)
Halley VI (UK)

Dome Fuji (Japan)

Mawson (Australia)
Druzhnaya (Russia)
Progress (Russia)
Zhongshan (China)
Davis (Australia)
Law (Australia)
Bharati (India)

Sky Blu (UK)

Taishanm (China)
Kunlun (China)

Parodi (Chile)

Amundsen–Scott South Pole Station (USA)

Mirny (Russia)

Vostok (Russia)

McMurdo (USA)

Concordia (France/Italy)

Casey (Australia)

Scott Base (New Zealand)
Mario Zucchelli (Italy)
Jang Bogo (Korea)

Dumont d'Urville (France)

MAXIMUM SUMMER POPULATION

1–20  20–50  50–100  100–200  200–500  500+

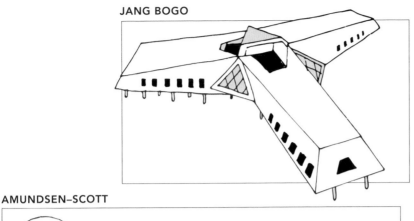

JANG BOGO

AMUNDSEN–SCOTT

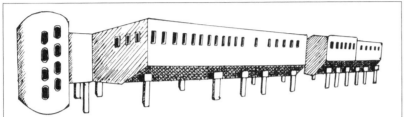

TAISHAN

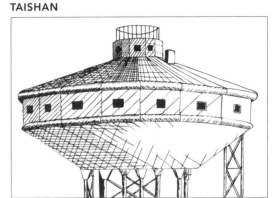

### AMUNDSEN-SCOTT *(USA)*

Amundsen–Scott Station, located at the Geographic South Pole, is the second largest station on the continent, and, with a floor area of 7,600 square metres, it is almost certainly the largest single building in Antarctica. It can accommodate 250 personnel with state-of-the-art laboratories, workshops and living space, all of which are raised above the surface of the ice by thirty-five hydraulic legs. It has been designed to copy the profile of an aeroplane wing, so that snow flows freely over and around the buildings and minimizes accumulation.

### JANG BOGO *(South Korea)*

The massive, space-age structure of Jang Bogo station is one of the most recently completed Antarctic bases, only finished in 2017. The base is located in Terra Nova Bay, on the Ross Sea, close to the Italian Zucchelli station. The three-pronged main building is just one of sixteen buildings with a further twenty-four observation facilities making this station one of the largest facilities other than the massive American stations. It is the second year-round Korean base, showing Korea's long-term commitment to Antarctic research.

### NEUMAYER III *(Germany)*

Like its British counterpart, Halley, the German research facility of Neumayer, run by the Alfred Wenger Institute, is situated on a floating ice shelf and has had several reincarnations. But, as the Ekström Ice Shelf moves slower than the Brunt, this is Neumayer's third building compared to Halley's sixth. The type of science conducted on the base is also similar, although as well as focusing on atmospheric research, Neumayer also acts as a logistic base for the German Antarctic programme. The building itself stands on legs six metres above the ice and rises to almost 30 metres. Under the structure is a subterranean garage, calved from the ice.

### TAISHAN *(China)*

It looks like an odd flying saucer, but China's Taishan research station is cleverly designed to channel wind around the building and avoid snow accumulation. The illustration shows the largest of the seven buildings on the base, which is located in a remote part of East Antarctica, halfway between two of China's other stations: Zongshan on the coast and Kunlun, high in the polar interior. Taishan acts as a staging post on the route between them.

**PRINCESS ELISABETH**

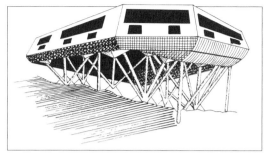

**CONCORDIA**

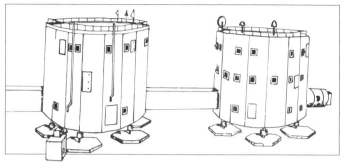

**HALLEY VI**

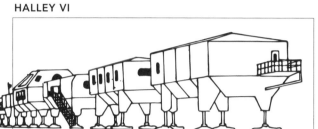

**BHARATI**

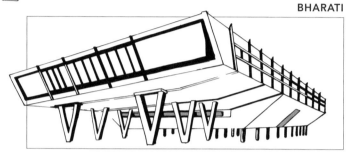

**NEUMAYER III**

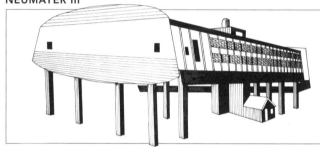

## BHARATI *(India)*

India's third research station, Bharati, is a science base focused on oceanography and geological studies. It is located close to three other international stations in the Larsemann Hills area of East Antarctica. From the station's sleek exterior design, you would not guess that it is designed around 134 interlocked shipping containers, surrounded by an insulating skin and a shiny silver outer shell.

## PRINCESS ELISABETH *(Belgium)*

Perched high on a cliff overlooking the remote ice fields of Dronning Maud Land, this sleek silver building looks like a perfect hideout for a Bond villain. But it is anything but. Surrounded by wind-turbines and solar panels, it is the only carbon-neutral base in Antarctica and is built of eco-friendly materials. It is energy-efficient and uses an innovative waste management system to process waste water. Belgium, which runs the station, may not be one of the largest nations in the Antarctic club, but this stunning, cutting-edge base is the envy of many.

NEUMAYER III

PRINCESS ELISABETH

HALLEY VI

BHARATI

TAISHAN

SCOTT–AMUNDSEN

CONCORDIA

JANG BOGO

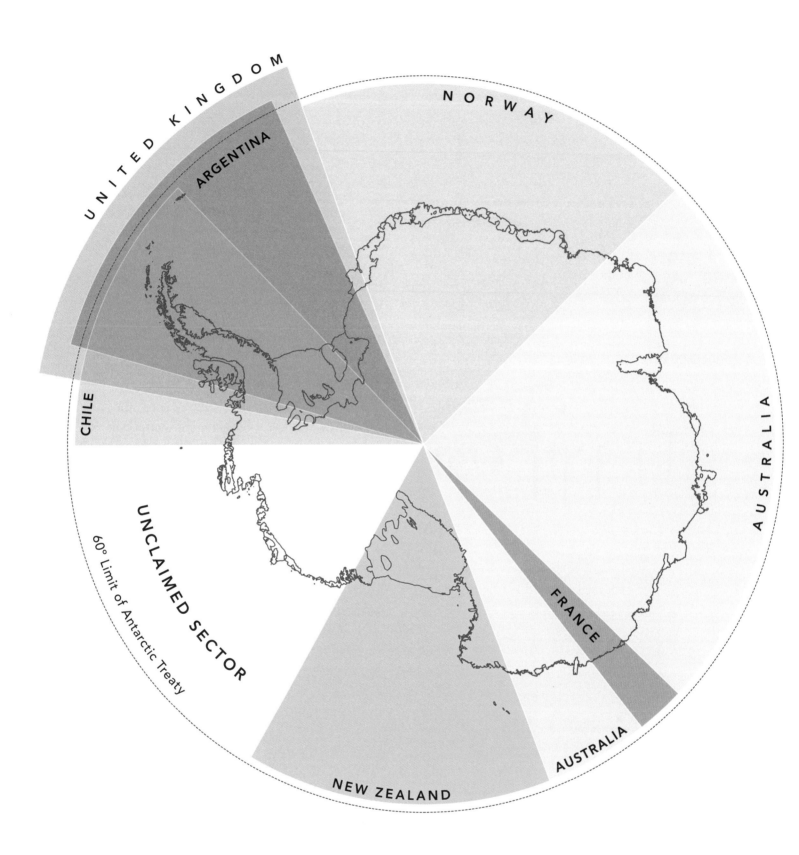

# 49. *Pieces of pie*

**BY THE LATE 1950S**, seven nations had put forward territorial claims to parts of the Antarctic continent, dividing it up into segments. The two superpowers of the day, the Soviet Union and the US, both refused to recognize any other claims and reserved the right to make future claims. Additionally, three of the claims – those of the British, the Argentinians and the Chileans – overlapped. Political tensions were building. Who would get access to the land and the potential mineral wealth of this vast new continent? Would this Cold War stand-off ferment into an even colder Antarctic war?

Something needed to be done, but the catalyst for change was not driven by politicians. In 1957–8, a surprising turn of events changed the future of the continent for ever. Scientists came together to collaborate in an international multidisciplinary project called the International Geophysical Year. Science knows no frontiers, and researchers from many different nations came together in Antarctica to cooperate on a variety of interdisciplinary projects to try to understand this mysterious new land. The collaboration was so successful that it gave the impetus needed for a political agreement: the Antarctic Treaty.

So what happened to the claims? Well, they are still there, put on hold while the treaty lasts. Governments often look at the long term in foreign affairs and want to keep their historical claims just in case. After all, who knows what the future will bring?

# 50. Who owns it?

**AFTER THE SUCCESS** of the International Geophysical Year in 1957–8 it was agreed that the continent of Antarctica would be run not as a nation state, but by a committee of interested governments. The principles of the agreement would be set down in the Antarctic Treaty.

The treaty banned militarization and nuclear weapons, ensuring that the continent is used only for peaceful purposes like scientific research, and allowing freedom of access to scientists of different nations. Later, in 1991, an environmental agreement was added to the treaty, prohibiting the exploitation of minerals and oil resources.

The twelve countries whose flags are depicted in the centre of the map signed the treaty in 1959, and it came into force in 1961. Today, fifty-three nations, including those whose flags are shown in the outer circle of the map, are signatories to the treaty.

The governments of these nations meet each year to discuss and agree on points of governance such as environmental protection and good stewardship. The Antarctic Treaty System is not without its problems, but overall it can be held up as a model of how nations can work together to govern a continent for peace and science.

## BUILDINGS

1. Flammables store
2. Vehicle store
3. Waste management
4. Garage
5. Cold storage
6. Fuel depot
7. Administration offices
8. Post office
9. Greenhouse
10. McMurdo Station Core Facility
11. Hospital
12. Bar
13. Club
14. Gym
15. Firehouse
16. Joint Spacecraft Operations Center
17. Café
18. Church
19. Power plant
20. Waste water plant
21. Water plant
22. Mac. ops communications
23. Cary Science & Engineering Center
24. NSF building
25. NSF dorm
26. Berg Field Center
27. Gym
28. Helicopter hangar
29. Helipad
30. Maintenance construction & craft center
31. Science support center
32. Old power plant
33. Electrical warehouse
34. Discovery hut
35. Hut Point Memorial

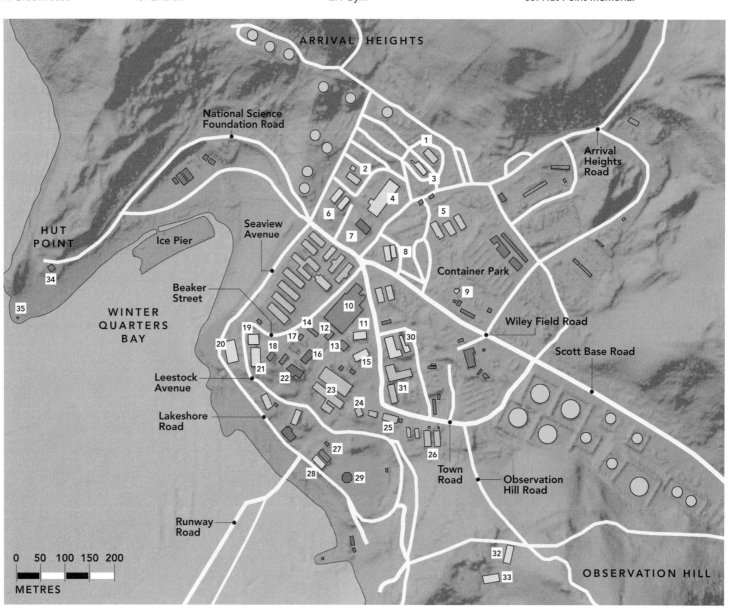

## KEY

- 🔵 Fuel tank
- ▫️ Storage
- ⬜ Science facility
- ⬛ Administration
- ⬜ Technical facility
- ⬛ Recreation & leisure
- ⬜ Accommodation
- ⬛ Not recorded
- 🔴 Helipad
- ▬ Road

# 51. Mac Town

**McMURDO STATION**, often shortened to McMurdo or Mac Town by its residents, is the American research station on Ross Island, and is by far the largest settlement in Antarctica. More like a small town than a research station, McMurdo boasts a cinema, coffee shops, bars, a club, a gym and several restaurants. It even has its own bus station and taxi service. McMurdo clings to the southern shore of Hut Point, the historic peninsula where both Scott and Shackleton overwintered before attempting their conquest of the Pole. McMurdo was first established as a US base in 1955, making it one of the oldest stations on the continent, and has been continuously occupied ever since. At maximum capacity, it can house around 1,300 people, although its more regular summer population is around a thousand, which makes it at least twice as large as any other Antarctic research station.

The main purpose of the base is to serve as a logistics and science hub for the US Antarctic Program. From here, the Program sends out planes, tractor-trains and helicopters to supply its other bases and remote field parties around the continent. The station has two ice runways, one located on sea ice for smaller planes, and one built on the more solid ice shelf for larger, intercontinental jets. Most of the cargo and fuel for the base is supplied by ship. Because it is so far south, even at the height of summer the sea is often frozen, and every year a channel has to be cut through the ice by powerful icebreakers in order to access the land.

**McMURDO STATION**

# 52. Mobile home

**THE HALLEY VI** research station is one of only two permanent stations in Antarctica to be situated on a floating ice shelf. This brings some unique challenges. The conditions on the low, flat ice shelf are cold and windy with lots of snow. The Halley base has been located on the Brunt Ice Shelf since 1958 and is in its sixth reincarnation. The first three versions of Halley were regular wooden buildings. These were built to withstand the cold and wind, but the blowing snow built up against the sides until they were each, in turn, buried and crushed. The fourth Halley was designed and engineered to withstand huge pressures of snow and ice above, but as the buildings were buried deeper and deeper, even the strengthened superstructure of the rooms and corridors squashed and buckled until this station, too, was uninhabitable. Halley V was raised off the ground on legs, which were jacked up every year to solve this problem. This way, the snow blew

underneath and didn't build up against the base. But there is another problem with locating a station on a floating ice shelf: the ice on the shelf is continually moving towards the sea. This movement is faster on ice shelves, often ten times as fast as ice movement on the mainland. Eventually, the shelf ice reaches the sea and breaks off into icebergs. At Halley, the base moves seaward by about a kilometre a year, so after several decades on the shelf Halley V started to near the coastal calving zone where the icebergs break off – it was time to build another base.

Halley VI was a radical new design. To get around the problem of the moving ice shelf the new research station was designed to be broken apart and towed away from the coast every few years. The station consists of eight pods, each standing on

*(Continued overleaf)*

## STATION MODULES

**H2 SCIENCE MODULE 2**

Metrology observation platform, ozone monitoring, workshops, dry lab.

**E2 ENERGY MODULE 2**

Generators, fuel, plumber, sewage treatment, boiler room.

**E1 ENERGY MODULE 1**

Generators, fuel, water treatment plant, boiler room, fire suppression unit.

**C CONTROL MODULE**

Comms, base commander, doctor, laundry, servers, boot room.

**B2 BEDROOM MODULE 2**

Bunks, storage.

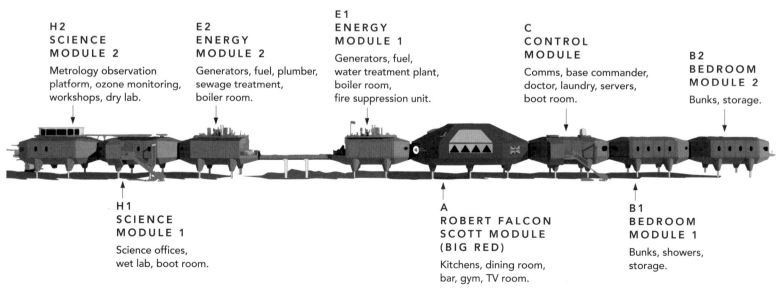

**H1 SCIENCE MODULE 1**

Science offices, wet lab, boot room.

**A ROBERT FALCON SCOTT MODULE (BIG RED)**

Kitchens, dining room, bar, gym, TV room.

**B1 BEDROOM MODULE 1**

Bunks, showers, storage.

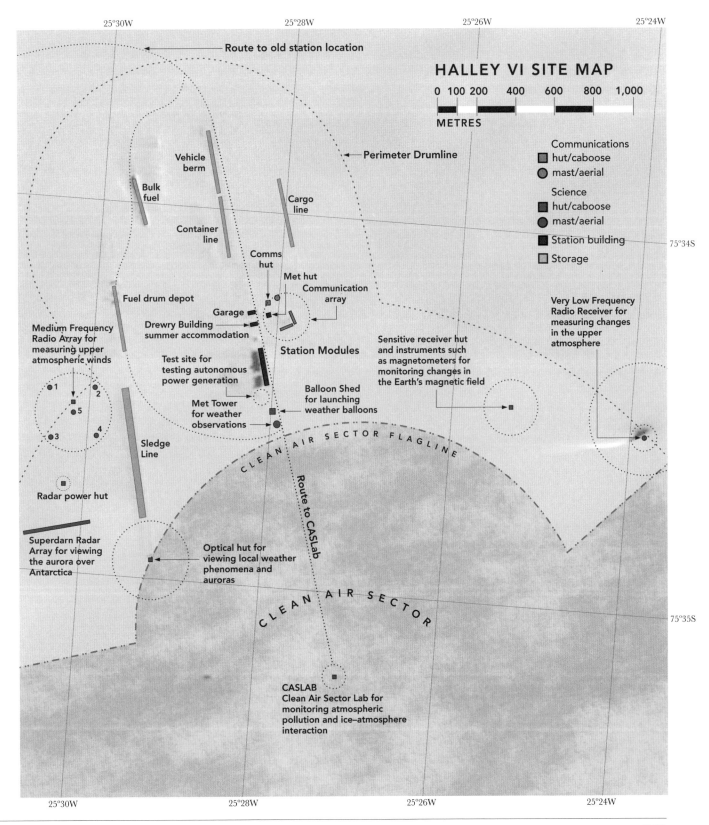

**HALLEY VI SITE MAP**

0  100  200  400  600  800  1,000
METRES

Route to old station location

Perimeter Drumline

Communications
☐ hut/caboose
◉ mast/aerial
Science
■ hut/caboose
● mast/aerial
■ Station building
☐ Storage

Vehicle berm

Bulk fuel

Container line

Cargo line

Comms hut

Met hut

Communication array

Fuel drum depot

Garage

Drewry Building summer accommodation

Medium Frequency Radio Array for measuring upper atmospheric winds

Test site for testing autonomous power generation

Station Modules

Very Low Frequency Radio Receiver for measuring changes in the upper atmosphere

Sensitive receiver hut and instruments such as magnetometers for monitoring changes in the Earth's magnetic field

Balloon Shed for launching weather balloons

1
2
5
3
4

Met Tower for weather observations

Sledge Line

Radar power hut

CLEAN AIR SECTOR FLAGLINE

Superdarn Radar Array for viewing the aurora over Antarctica

Route to CASLab

Optical hut for viewing local weather phenomena and auroras

CLEAN AIR SECTOR

CASLAB
Clean Air Sector Lab for monitoring atmospheric pollution and ice–atmosphere interaction

25°30W    25°28W    25°26W    25°24W

75°34S

75°35S

0    5    10    15    20

**KILOMETRES**

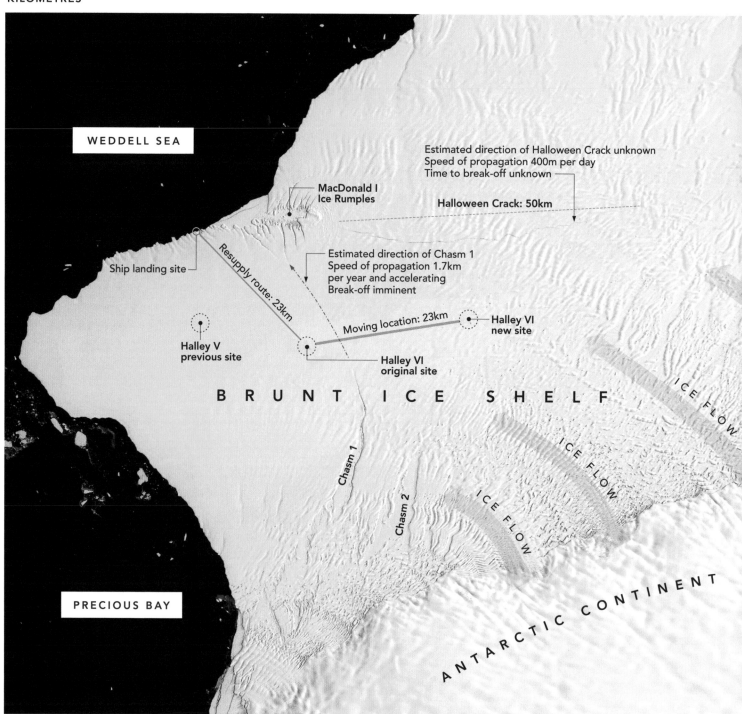

WEDDELL SEA

Estimated direction of Halloween Crack unknown
Speed of propagation 400m per day
Time to break-off unknown

**Halloween Crack: 50km**

MacDonald I
Ice Rumples

Ship landing site

Estimated direction of Chasm 1
Speed of propagation 1.7km
per year and accelerating
Break-off imminent

Resupply route: 23km

Moving location: 23km

Halley VI
new site

Halley V
previous site

Halley VI
original site

B R U N T    I C E    S H E L F

ICE FLOW

Chasm 1

Chasm 2

ICE FLOW

ICE FLOW

ICE FLOW

PRECIOUS BAY

ANTARCTIC CONTINENT

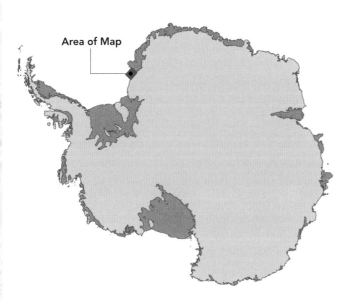

**Area of Map**

four legs (which can be decoupled from each other). Each of the legs, critical to stop snow build-up, has a giant ski on the bottom, which allows the modules to be pulled by tractors. The new base was opened in 2012.

The movement of the base happened sooner than expected. Soon after completion of the station, a giant chasm that had been dormant for years started to widen. The end of the crack was cutting across the shelf, extending at a rate of around a metre a day. Projections estimated that if this

massive crack continued to widen it would cut across the Brunt Ice Shelf, breaking it in two and creating an enormous iceberg on the seaward side. Unfortunately, the new Halley VI buildings were on the wrong side of the chasm and would float off on the berg when it calved. The decision was made to move the station. In early 2017 the pods were broken apart and pulled by a fleet of powerful tractors a distance of 23 kilometres, away from the coast to a new, safer location.

And that should have been the end of the story, but the Brunt Ice Shelf had other ideas. On 31 October 2016, just as the station was being prepared to be moved, another, unforeseen crack appeared to the north of the station and started to rip across the ice shelf at an alarming rate. Within a month, this Halloween Crack had grown to 20 kilometres in length. Glaciologists realized that they could not safely predict the behaviour of the ice shelf and so, for the first time in its history, it was decided that Halley would not operate in the winter, when station staff are isolated and cannot be rescued, and would, until conditions become more predictable, be used as a summer-only station. The moving of Halley was a superbly successful example of planning and logistics, but Antarctica, as ever, has had the last word.

# 53. International antics

**MANY NATIONS BUILD** their bases on the small archipelago called the South Shetland Islands. From here, northwards across Drake Passage, South America is a mere 1,000 kilometres away, making these islands the closest part of Antarctica to the outside world. These islands also have the advantage of having a milder climate than elsewhere on the continent and are free of sea ice for most of the year. There are abundant areas of bare rock on which to build a research station, a condition that is not common in other parts of the frozen wilderness. These advantages mean that the central portion of this small group of islands has more research stations, from more countries, than anywhere else on the continent. From Peru to Poland, Brazil to Bulgaria, it seems like you are just not in the Antarctic club unless you have a base here. Twenty-one research stations from thirteen countries occupy the archipelago, with the greatest concentration around Maxwell Bay on King George Island. Here, the Chileans have a gravel airstrip that serves many of the surrounding stations as a logistics hub. Indeed, it is here that each Antarctic winter, scientists and support staff from many countries gather together to contest the 'Antarctic Olympics'. However, the events aren't the typical Olympic sports such as athletics, swimming or cycling. Bearded overwinterers who have honed their skills in the type of games that can be played in confined research stations over the long darkness of the Antarctic winter compete in table football, ping-pong, darts, pool and poker, as well as the slightly more traditional sports of badminton and volleyball. Rivalries between the bases ensure that competition is fierce, with national pride at stake. Woe betide if the Chinese lose the ping-pong or the Russians fail to win the volleyball!

**BASES**

1. International Field Camp
2. Guillermo Mann
3. Ohridski
4. Juan Carlos I
5. Decepción
6. Gabriel de Castilla
7. Cámara
8. Risopatrón
9. Maldonado
10. Arturo Prat
11. Julio Escudero
12. Bellingshausen
13. Great Wall
14. Artigas
15. King Sejong
16. Dallman
17. Carlini
18. Machu Picchu
19. Arctowski
20. Comandante Ferraz
21. Refugio Ecuador

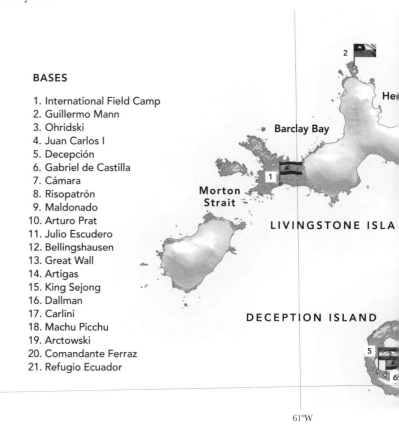

63°S

61°W

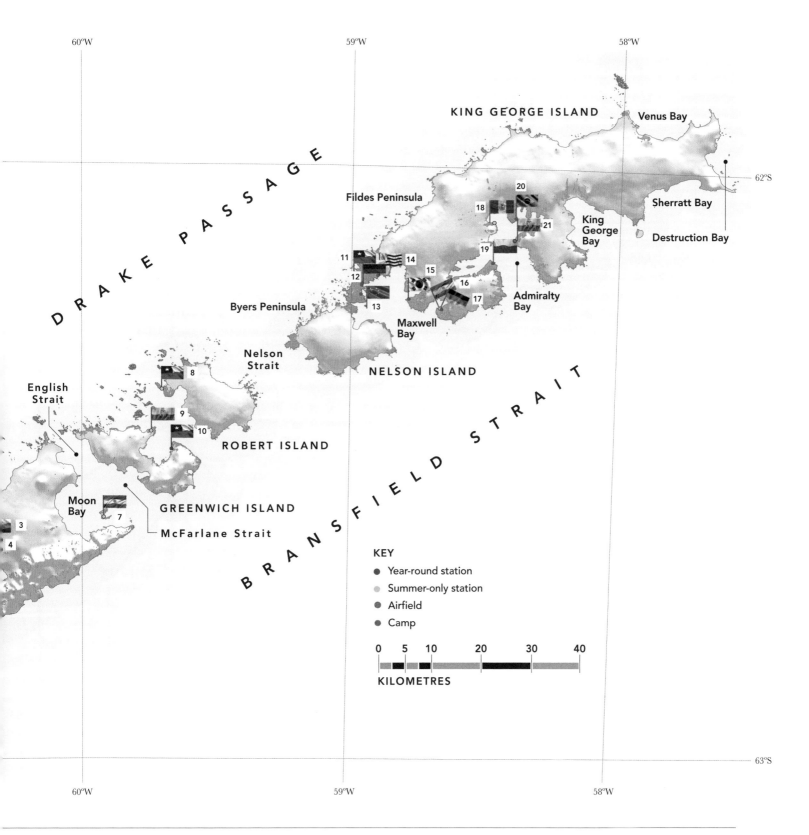

60°W

59°W

58°W

**DRAKE PASSAGE**

**KING GEORGE ISLAND**

Venus Bay

62°S

**Fildes Peninsula**

Sherratt Bay

20

18

21

King
George
Bay

Destruction Bay

11

14

19

12

15

16

**King
George
Bay**

Admiralty
Bay

**Byers Peninsula**

13

17

**Maxwell
Bay**

**NELSON ISLAND**

**Nelson
Strait**

8

**BRANSFIELD STRAIT**

**English
Strait**

9

10

**ROBERT ISLAND**

**GREENWICH ISLAND**

**Moon
Bay**

7

3

**McFarlane Strait**

4

**KEY**

● Year-round station

● Summer-only station

● Airfield

● Camp

| 0 | 5 | 10 | 20 | 30 | 40 |
|---|---|----|----|----|----|

**KILOMETRES**

60°W

59°W

58°W

63°S

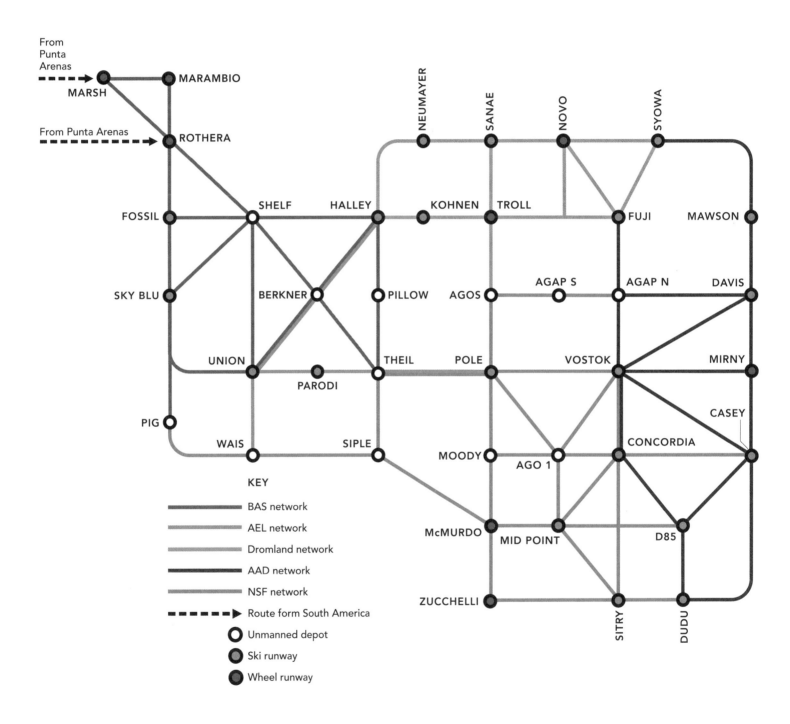

From
Punta
Arenas

MARSH

MARAMBIO

From Punta Arenas

ROTHERA

NEUMAYER

SANAE

NOVO

SYOWA

FOSSIL

SHELF

HALLEY

KOHNEN

TROLL

FUJI

MAWSON

SKY BLU

BERKNER

PILLOW

AGOS

AGAP S

AGAP N

DAVIS

UNION

PARODI

THEIL

POLE

VOSTOK

MIRNY

PIG

CASEY

WAIS

SIPLE

MOODY

AGO 1

CONCORDIA

McMURDO

MID POINT

D85

ZUCCHELLI

SITRY

DUDU

KEY

BAS network

AEL network

Dromland network

AAD network

NSF network

Route form South America

Unmanned depot

Ski runway

Wheel runway

# 54. Antarctic skies

**THERE ARE NO ROADS** in Antarctica. Travel over the featureless ice sheet in the interior of the continent is difficult and dangerous. Extreme conditions and hidden crevasses await the overland voyager, and almost all travel of any distance is by aircraft. A network of flight routes has developed between airstrips to connect the disparate research stations, but few hard runways exist and most of the landing strips are on ice so the planes have to land and take off on skis.

Two types of aircraft dominate the internal flight network, the Basler BT-67 and the de Havilland Twin Otter. The Basler is a turbo-prop version of an old Douglas DC-3, a design that dates back to the 1930s, while the smaller Twin Otter, also an old design, was developed in the 1960s. Both are rugged yet reliable, with short take-off and landing capabilities. However, neither has a range of more than 1,000 nautical miles. This means that the only route into Antarctica is via the shortest crossing of the Southern Ocean, across Drake Passage from South America to Marsh or Rothera station on the Antarctic Peninsula. From there, the planes must make their way down the Peninsula and around the flight network, refuelling multiple times, to reach their final destination.

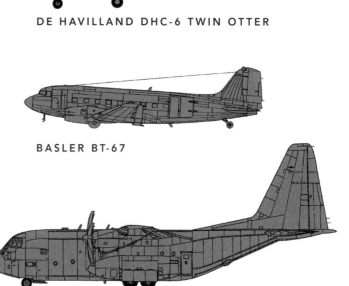

DE HAVILLAND DHC-6 TWIN OTTER

BASLER BT-67

LOCKHEED C-130 HERCULES

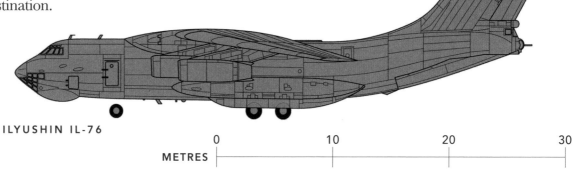

ILYUSHIN IL-76

METRES  0   10   20   30

# 55. *Exploiting the ocean*

THE SOUTHERN OCEAN is rich in marine life but, due to its remoteness and rough conditions, fishing in it is expensive and dangerous. However, as other fish populations around the globe have been overexploited and devastated, the pressure to catch the fish and krill of Antarctica has grown. In the 1970s, toothfish around parts of the continent were drastically overfished and their populations plummeted. So, in 1982, the international community set up the Commission for the Conservation of Antarctic Marine Living Resources (usually known by its acronym CCAMLR) to protect and conserve the ecology and wildlife of the Southern Ocean. CCAMLR sets quotas for how many and where different species of fish or krill can be caught, dividing the ocean into a number of zones. The commission takes a precautionary approach to fishing quotas, trying to ensure that the levels of commercial catches have no knock-on consequences for the huge populations of predators, such as whales,

seals and penguins which also rely on the living resources of the oceans around Antarctica.

These zones are shown on the map and colour coordinated depending upon what combination of resources may be taken from each area. The main target species for the fishermen is the large and lucrative Antarctic toothfish, which can grow up to 1.7 metres in length and weigh as much as 160 kilograms. The second most important resource taken from the ocean is Antarctic krill, which is quite literally hoovered out of the ocean by large factory-fishing ships. Most krill is used for animal feed, but a small percentage of the best catch is utilized in the health food and pharmaceutical industries (see Keystone Krill map, page 108). Although the value of krill taken is tiny compared to that of toothfish, the volume taken is enormous, sometimes exceeding 100,000 tonnes per year.

Toothfish

Antarctic Krill

Icefish

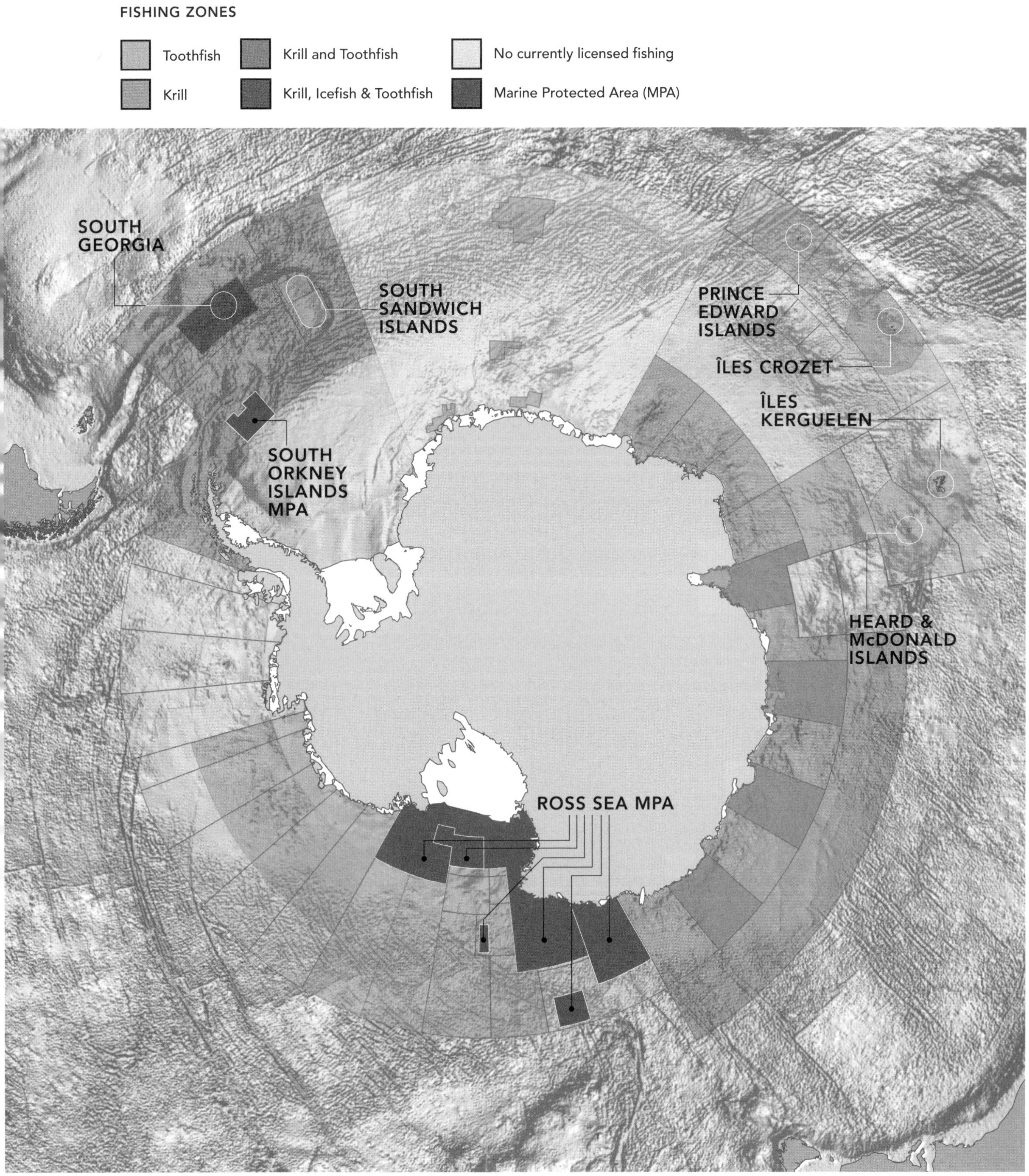

**FISHING ZONES**

- Toothfish
- Krill
- Krill and Toothfish
- Krill, Icefish & Toothfish
- No currently licensed fishing
- Marine Protected Area (MPA)

SOUTH
GEORGIA

SOUTH
SANDWICH
ISLANDS

PRINCE
EDWARD
ISLANDS

ÎLES CROZET

ÎLES
KERGUELEN

SOUTH
ORKNEY
ISLANDS
MPA

HEARD &
McDONALD
ISLANDS

ROSS SEA MPA

# 56. Tourist hub

**TOURISM IN ANTARCTICA** is becoming big business. Each year, around 40,000 tourists visit the icy shores, drawn by the penguins, the scenery and the continent's sheer other-worldliness. Almost all of them come on cruise ships and visit the Antarctic Peninsula region. This area has the greatest diversity of wildlife and some of the best and most accessible scenery in Antarctica. It is also the region that has the least sea-ice cover and is therefore the safest for tourist ships to operate in. Most of them will anchor at a limited number of specific locations, usually penguin colonies, historic sites such as monuments dedicated to explorers of the past, research stations or scenic spaces, while navigating the celebrated straits and waterways of the Peninsula, such as the stunning Lemaire and Neumayer channels.

The map shows the density of cruise ships in Antarctica each year, indicating the typical paths they follow, as well as the top twenty-five tourist sites on the continent. Some of these sites, such as Deception Island, are in the South Shetland archipelago, but the vast majority are in a small area between the Gerlache Strait and Grandidier Channel. This is a beautiful region, with impressive bays, many penguin colonies and sheer mountains that rise straight from the crystal-clear waters of the Antarctic Peninsula. Visiting Antarctica is never cheap, but the magical scenery, the wildlife and the environment combine to provide an experience that few people ever forget.

## TOP 25 VISITOR SITES

1. Neko Harbor
2. Cuverville Island
3. Goudier Island
4. Half Moon Island
5. Whalers Bay
6. Petermann Island
7. Almirante Brown Station
8. Jougla Point
9. Danco Island
10. Brown Bluff
11. Vernadsky Station
12. Telefon Bay
13. Barrientos Island
14. Orne Harbor
15. Yankee Harbor
16. Mikkelsen Harbor
17. Damoy Point
18. Paradise Bay
19. Pléneau Island
20. Hannah Point
21. Port Charcot
22. Great Wall Station
23. Yalour Islands
24. Waterboat Point
25. Bellingshausen Station

## KEY

 Penguin colony

 Historic site or monument

 Scenic site

 Research station

 Scenic ship passge

## TOURIST SHIPS PER YEAR

 1–5

 6–20

 21–100

100+

65°W

60°W

62°S

S O U T H   S H E T L A N D   I S L A N D S

King George Island

HS
M

25

22

13

15

Livingstone
Island

20

4

12

5

Deception
Island

HS
M

HS
M

Smith Island

Low Island

BRANSFIELD STRAIT

Antarctic Sound

Joinville
Island

10

Terror and Erebus Gulf

Trinity
Island

Gerlache Strait

Brabant Island

16

James Ross Island

Seymour Island

64°S

Anvers Island

HS
M

Snow Hill Island

HS
M

2   14

9

17

24

3   8

7   18   1

T R I N I T Y
P E N I N S U L A

HS
M

21

19

23

6

11

Larsen Island

W E D D E L L
S E A

Grandidier
Channel

HS
M

Renaud Island

Crystal Sound

G R A H A M   L A N D

Adelaide
Island

66°S

L A R S E N   C   I C E   S H E L F

0    25    50    75    100

KILOMETRES

Marguerite Bay

65°W

60°W

55°W

# CHAPTER 8

# *Exploration*

57. To find a continent *152*

58. The heroic age *154*

59. The greatest escape *156*

60. The race that never was *160*

61. Get on your knees and pray *164*

62. The home of the blizzard *168*

63. Exploring from above *172*

64. Postwar power play *174*

65. The scientific age *176*

66. The satellite age *178*

67. A most historic place *180*

68. Traces of the past *182*

# 57. *To find a continent*

**YOU WOULD BE FORGIVEN** for assuming that finding a continent twice the size of Australia would be easy, but that was not the case with Antarctica. Captain Cook, one of history's greatest navigators, tried circumnavigating the Southern Ocean twice in the 1770s, but he never set eyes on land, unable to penetrate the dense barrier of pack ice that surrounds the coast. From the mountainous icebergs he sighted he surmised that land must exist further to the south but that it would be so desolate it would not be worth the effort of finding it. He was right that land existed, but he was wrong about its value.

It was another fifty years before Antarctica was eventually sighted, and then, by three explorers almost in the same year. Exactly who 'discovered' the continent is a matter of debate among scholars, and that debate revolves around exactly what constitutes Antarctica.

The first claim is by William Smith, the British captain of a freighter who was on his way around Cape Horn to Valparaiso. Smith was running south in Drake Passage to avoid one of the devastating storms that are frequent in that part of the world. As dawn broke on 19 February 1819, he saw the tall, rocky island that now bears his name. The island is part of the South Shetland Islands, the northernmost archipelago in Antarctica. Closer inspection showed that the coast played host to many fur seals, a potential goldmine for sealers, and Smith knew it. He sailed on to Chile, eager to keep his discovery secret until he could inform the British Admiralty office in Santiago. When he arrived there, the Admiralty commandeered his ship and placed a navy captain, Edward Bransfield, in charge. As soon as spring arrived, they sailed back to the newly discovered islands. What a shock they must have had! By the time they got there, several other ships were already traversing the island chain. They belonged to sealers. William Smith may not have given the game away, but his sailors had gossiped in port, and the seal hunters, knowing the value of pristine hunting grounds, had sailed as soon as they heard the news. Captain Bransfield landed and claimed the territory for Britain on 2 February 1820. He pushed further south and discovered Trinity Peninsula, the northern part of the Antarctic mainland, but did not land there. Nathaniel Palmer, an American sealer, has the honour of being the first to set foot on the mainland, on 17 November 1820.

Another captain could, however, claim to have seen the mainland before him. The Russian Fabian von Bellingshausen was circumnavigating Antarctica around the same time. Like Cook, he had been sent on an official expedition to find southern lands, in his case by the Russian government. Also like Cook, he came close to land on many occasions, only to be beaten by the pack ice. On 27 January 1820, he spotted a high wall of ice on the southern horizon. Today, we know that the position he recorded is near the terminus of the Fimbulisen Ice Tongue, the large, floating extremity of one of Antarctica's great ice shelves. He had probably seen that part of Antarctica, but this barren ice was not what he was hoping for, and it lent weight to Cook's claim that any land would be worthless. Von Bellingshausen had seen an ice shelf – but was it land?

So we have four contenders for discoverer of Antarctica: Smith, who saw the outlying islands; Bellingshausen, who saw the ice shelf; Bransfield, who sighted the mainland and Palmer, who landed on it. I will leave it to you to decide who found the continent of Antarctica.

Over the following twenty years, the rich bounty of seals around the coast and the search for new hunting grounds opened up much of the coastline. Many sealing captains and explorers penetrated the pack ice, rapidly mapping the coastline around the continent. But by 1840 the seals had been annihilated and so the commercial interest, the very reason for exploration, was lost.

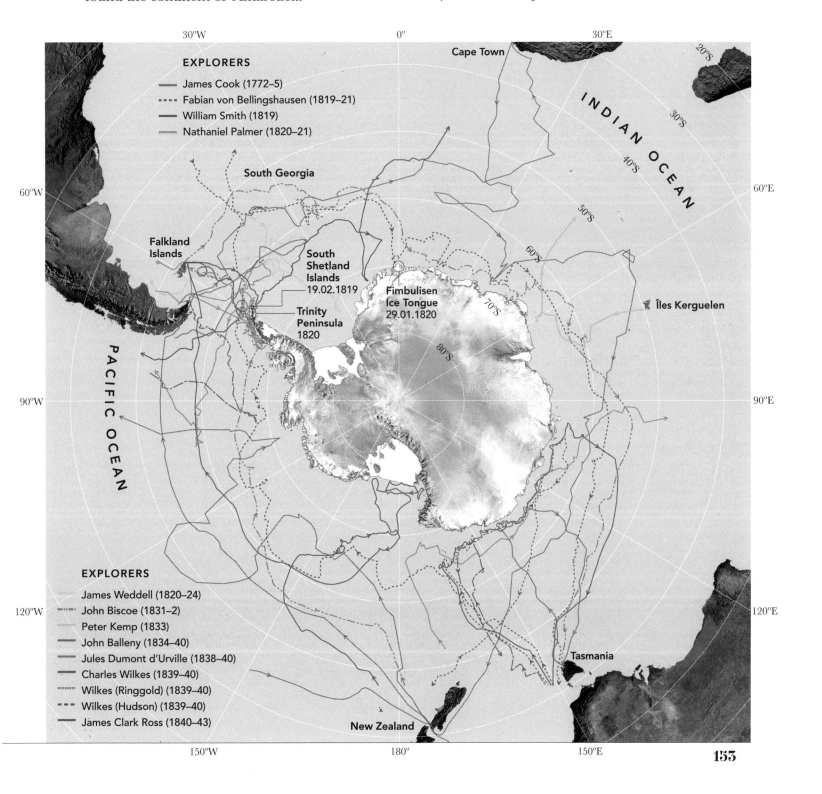

**EXPLORERS**
— James Cook (1772–5)
···· Fabian von Bellingshausen (1819–21)
— William Smith (1819)
— Nathaniel Palmer (1820–21)

Cape Town

INDIAN OCEAN

20°S
30°S
40°S
50°S
60°S

South Georgia

Falkland Islands

South Shetland Islands 19.02.1819

Trinity Peninsula 1820

Fimbulisen Ice Tongue 29.01.1820

70°S
80°S

Îles Kerguelen

PACIFIC OCEAN

**EXPLORERS**
— James Weddell (1820–24)
···· John Biscoe (1831–2)
— Peter Kemp (1833)
— John Balleny (1834–40)
— Jules Dumont d'Urville (1838–40)
— Charles Wilkes (1839–40)
···· Wilkes (Ringgold) (1839–40)
···· Wilkes (Hudson) (1839–40)
— James Clark Ross (1840–43)

Tasmania

New Zealand

150°W
180°
150°E

30°W
0°
30°E
60°W
60°E
90°W
90°E
120°W
120°E

## EXPLORERS

- Adrien de Gerlache (1898–9)
- Carsten Borchgrevink (1898–1900)
- Enrich von Drygalski (1902–3)
- Otto Nordenskjöld (1902–3)
- Robert Scott (1902–4)
- William Bruce (1903–4)
- Jean Charcot (1908–10)
- Ernest Shackleton (1908–9)
- Robert Scott (1910–13)
- Roald Amundsen (1910–12)
- Wilhelm Filchner (1911–12)
- Douglas Mawson (1911–14)
- Ernest Shackleton (1914–16)

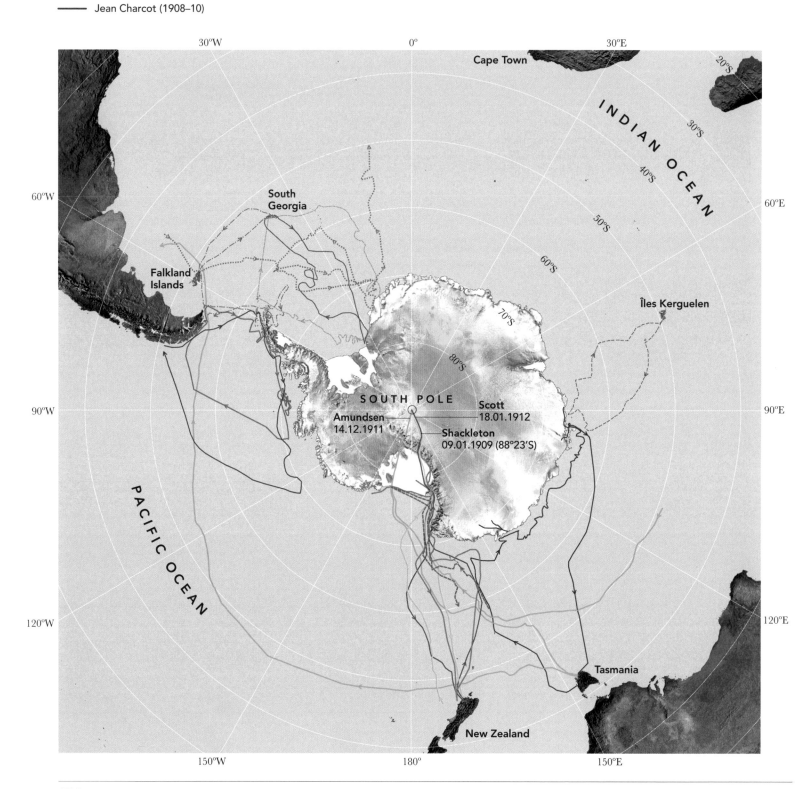

Cape Town

INDIAN OCEAN

20°S
30°S
40°S
50°S
60°S
70°S
80°S

South Georgia

Falkland Islands

Îles Kerguelen

SOUTH POLE

Amundsen
14.12.1911

Scott
18.01.1912

Shackleton
09.01.1909 (88°23'S)

PACIFIC OCEAN

Tasmania

New Zealand

30°W    0°    30°E
60°W    60°E
90°W    90°E
120°W    120°E
150°W    180°    150°E

# 58. *The heroic age*

**THE PERIOD WE NOW KNOW** as the 'Heroic Age of Exploration' had its origins in 1895. The mid-nineteenth century had seen a lull in the zeal to explore the southern continent. It was at the Sixth International Geographical Congress in London that a renewed interest in Antarctica was sparked. The meeting recommended geographic and scientific exploration of the frozen continent in the belief that this was the greatest remaining frontier of human investigation.

Explorers around the world rallied to the cause, some funded by governments or scientific institutions, others using their personal finances or relying on private donations. The Belgian de Gerlache was the first to overwinter, stuck in the sea ice west of the Peninsula in 1898, whilst the Norwegian Carsten Borchgrevink claimed the first landing in East Antarctica and also overwintered in 1899. Larsen, Nordenskjöld (see pages 156–9), Mawson (see pages 168–71) and others battled the elements in search of discovery on the fringes of the frozen continent, but it was the South Pole that was the greatest prize.

Scott was the first to attempt the heroic journey on the Discovery expedition of 1902–3. His team set up the first of many bases on Ross Island, at the edge of the Ross Ice Shelf, exploring the region and making important scientific discoveries. Scott's advance party found a route through the Transantarctic Mountains on to the polar plateau and attempted to reach the Pole, but turned back at just south of 82°S, still almost 800 kilometres from it.

Ernest Shackleton led the Nimrod expedition to Antarctica in 1908–9. He had been Scott's third officer on the Discovery expedition and, benefiting from that previous experience, his party forced a way through the Transantarctic Mountains to reach within 180 kilometres of the Pole. His attempt, which came so close to achieving its objective, proved that the Pole was attainable and was followed soon afterwards by the famous expeditions of Scott and Amundsen in 1911–12, which led to success for one party and tragedy for the other (see The Race That Never Was on pages 160–63).

With the Pole reached, the last objective was a complete crossing of the continent. This was the aim of Shackleton's ill-fated Imperial Trans-Antarctic expedition (see pages 164–67), which led to his crew's ultimate epic escape from Antarctica in 1916. This journey ended the Heroic Age. The world had been plunged into the horrors of the First World War, and it would be more than a decade before Antarctic exploration was contemplated once more.

# 59. *The greatest escape*

CARL ANTON LARSEN was a Norwegian sealer who became one of the most celebrated ship captains of what has become known as the 'Heroic Age of Exploration'. In 1893, he led an expedition down the eastern side of the Antarctic Peninsula. The sea ice that year was unusually light, and Larsen, on his ship, the *Jason*, pushed far into the Weddell Sea. His discoveries were later named in his honour – the Larsen Ice Shelf and Jason Peninsula. It would be almost a century before any other mariner travelled so far south in these waters. Encouraged by an account of the journey, Otto Nordenskjöld, a Swedish geologist and explorer, contracted Larsen to transport him and his small team to Antarctica. The Swede hoped to explore and overwinter somewhere in the area that Larsen had earlier discovered.

Initially, the journey went well and, in his ship, the *Antarctic*, Larsen passed south through Antarctic Sound, the iceberg-infested channel that now bears the ship's name. They pressed onwards until they could penetrate the ice no further, finally depositing Nordenskjöld's small party of six men on the northern tip of Snow Hill Island. The plan was for Larsen to collect the explorers during the Antarctic summer after they had overwintered, but the return trip did not go as planned. In December 1902, when Larsen reached Antarctic Sound it was choked with ice, and the ship could not get through it. Larsen knew that he would have to attempt the more treacherous route around Joinville Island, but if the sea-ice conditions were bad and he could not get the ship to Nordenskjöld's camp at Snow Hill Island, he would be unable to rescue the overwinterers. Before he departed Antarctic Sound, he left three men on the rocky outcrop of Hope Bay, telling them to sledge south to Snow Hill Island and return with Nordenskjöld and his men. The three set off but were soon thwarted by a tract of open water that blocked their path. Deflated, they returned to Hope Bay to wait for the ship.

Meanwhile, Larsen travelled east, rounding Joinville Island, and navigated his way south into Weddell Sea. The pack ice was thick and the ship was trapped on numerous occasions, eventually becoming stuck completely. The jammed hull was pinned and crushed by icebergs, it started to crack and take in water and, after some weeks, it broke up and sank. Salvaging what they could, the crew made their way across the shifting sea ice to the nearest land, the small, rocky outcrop of Paulet Island. There they built a hut from one of the lifeboats to try to survive the winter.

The adventurers were now in three groups – three at Hope Bay, nineteen on Paulet Island and the six scientists on Snow Hill Island, who had by now realized that they would have to face another Antarctic winter. For the two groups from the ship, the winter was terrible; having little food, they lived on penguins and seals. No one in the outside world knew whether they had survived or their whereabouts, and the possibility of rescue looked bleak. The small party at Hope Bay realized that their only hope of survival was to get to Nordenskjöld. At least the people back home knew where he was and that he needed rescuing. In the spring, they set off south once more, across the ice,

*(Continued overleaf)*

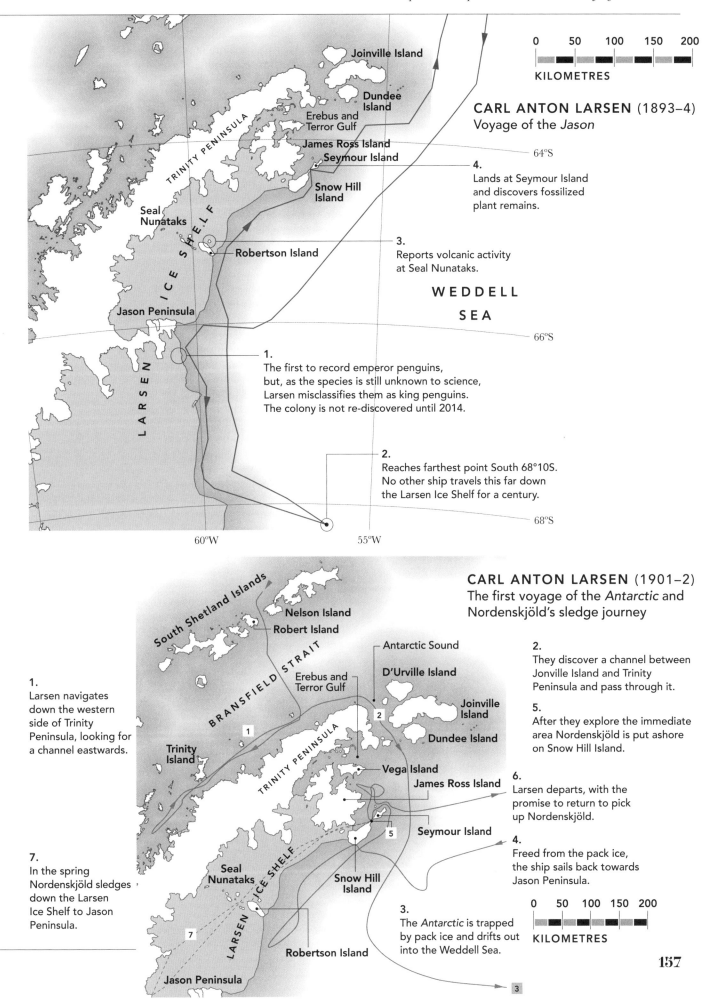

**CARL ANTON LARSEN** (1893–4)
Voyage of the *Jason*

Joinville Island

Dundee Island

Erebus and Terror Gulf

James Ross Island
Seymour Island

Snow Hill Island

TRINITY PENINSULA

Seal Nunataks

ICE SHELF

Robertson Island

LARSEN

Jason Peninsula

WEDDELL

SEA

64°S

66°S

68°S

60°W

55°W

**4.**
Lands at Seymour Island and discovers fossilized plant remains.

**3.**
Reports volcanic activity at Seal Nunataks.

**1.**
The first to record emperor penguins, but, as the species is still unknown to science, Larsen misclassifies them as king penguins. The colony is not re-discovered until 2014.

**2.**
Reaches farthest point South 68°10S. No other ship travels this far down the Larsen Ice Shelf for a century.

0   50   100   150   200
KILOMETRES

**CARL ANTON LARSEN** (1901–2)
The first voyage of the *Antarctic* and Nordenskjöld's sledge journey

South Shetland Islands

Nelson Island
Robert Island

BRANSFIELD STRAIT

Antarctic Sound

D'Urville Island

Joinville Island

Dundee Island

Erebus and Terror Gulf

Trinity Island

TRINITY PENINSULA

Vega Island

James Ross Island

Seymour Island

Seal Nunataks

ICE SHELF

Snow Hill Island

LARSEN

Robertson Island

Jason Peninsula

**1.**
Larsen navigates down the western side of Trinity Peninsula, looking for a channel eastwards.

**7.**
In the spring Nordenskjöld sledges down the Larsen Ice Shelf to Jason Peninsula.

**2.**
They discover a channel between Jonville Island and Trinity Peninsula and pass through it.

**5.**
After they explore the immediate area Nordenskjöld is put ashore on Snow Hill Island.

**6.**
Larsen departs, with the promise to return to pick up Nordenskjöld.

**4.**
Freed from the pack ice, the ship sails back towards Jason Peninsula.

**3.**
The *Antarctic* is trapped by pack ice and drifts out into the Weddell Sea.

0   50   100   150   200
KILOMETRES

**157**

## CARL ANTON LARSEN (1902–3)
The final voyage of the *Antarctic*

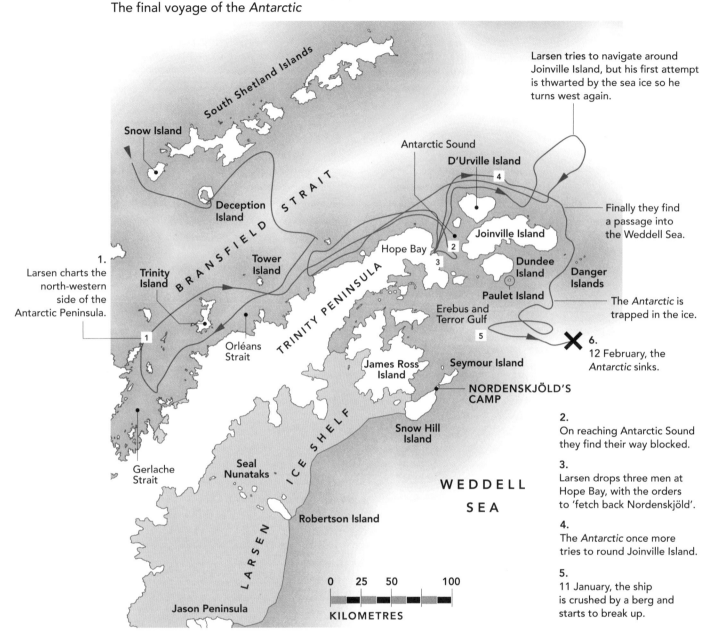

Larsen tries to navigate around Joinville Island, but his first attempt is thwarted by the sea ice so he turns west again.

**1.**
Larsen charts the north-western side of the Antarctic Peninsula.

Finally they find a passage into the Weddell Sea.

The *Antarctic* is trapped in the ice.

**6.**
12 February, the *Antarctic* sinks.

**2.**
On reaching Antarctic Sound they find their way blocked.

**3.**
Larsen drops three men at Hope Bay, with the orders to 'fetch back Nordenskjöld'.

**4.**
The *Antarctic* once more tries to round Joinville Island.

**5.**
11 January, the ship is crushed by a berg and starts to break up.

but they veered westwards towards Prince Gustav Channel. By sheer luck, it just so happened that Nordenskjöld, with two other men from his party, had decided to explore this channel at the same time. When they saw three wanderers in the distance they were so surprised they mistook them for giant penguins. As they got closer and realized they were human, they thought they had discovered 'native Antarcticans'. Imagine their surprise when they realized the three were part of the crew of the *Antarctic* – and what luck it was to

have found them!

Back in Sweden, a public petition had raised funds to charter a rescue ship. The Argentinians also sent a ship, the *Uruguay*, which navigated through Antarctic Sound towards Nordenskjöld. But what of Larsen? After wintering on Paulet Island, the captain chose five companions to row their remaining lifeboat across to Hope Bay to rescue their three stranded companions. However, on reaching the remains of their grimy hut, they found the three men gone, having left a note

## The Rescue of the Three Parties

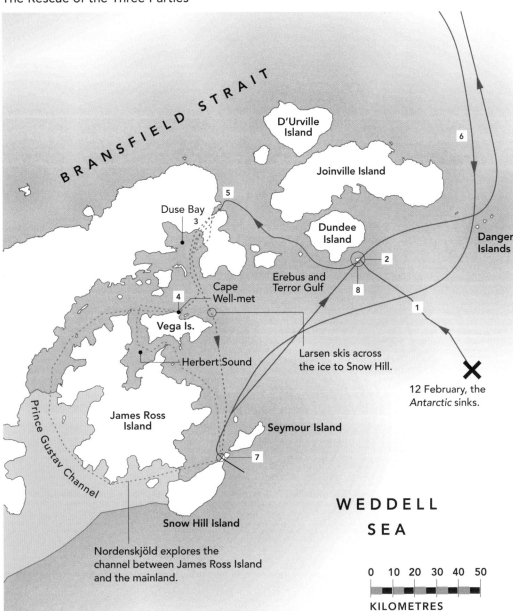

**1.**
By boat and foot Larsen and crew cross the pack ice to reach Paulet Island.

**2.**
Larsen's crew overwinter on Paulet Island, eating penguins to survive.

**3.** - - - - - - - - - - - - - -
The three-man party – Duse, Andersson and Grunden – after overwintering in a stone hut, set off southwards on skis to find Nordenskjöld.

**4.**
Nordenskjöld meets Duse's party on the ice, 12 October 1903.

**5.** ————————
In October 1902, Larsen and five men row a lifeboat to Hope Bay, only to find his men have already left.

**6.** ————————
Course of Captain Irizar and the *Uruguay*, 1903.

**7.**
8 November, men from the Argentinian rescue ship, *Uruguay*, reach Nordenskjöld's hut. By chance, Larsen arrives later that afternoon.

**8.**
8 November, the *Uruguay* rescues the remaining crew of the *Antarctic* from Paulet Island.

saying that they had set off to find Nordenskjöld. Larsen turned south once more to try to reach Snow Hill Island.

On 8 November 1903, the *Uruguay* reached Snow Hill Island and rescued Nordenskjöld's group and the three travellers from Hope Bay. Miraculously, on the very same day, Larsen arrived with his five-man party, frantically skiing to catch the ship they had seen in the distance. The rescue ship sailed northwards, collecting the remaining crew from Paulet Island on its voyage home. Against

all odds, everyone, even the ship's cat, had been rescued. All the men had survived, except for one crew member, who succumbed to illness while overwintering on Paulet Island.

*'The next moment, wild ear-piercing cheers mingled with shouts of "Larsen! Larsen is here!!" tear us away.'*

From Otto Nordenskjöld's account of their eventual meeting with Carl Anton Larsen at the end of their eventful expedition to overwinter on Snow Hill Island in 1903

Scott

Amundsen

## AMUNDSEN

| Bay of Whales | 80° South | 81° South | 82° South | 83° South | 84° South |
|---|---|---|---|---|---|
| 19 October | 23 October | 31 October | 5 November | 9 November | 13 November |

ROSS ICE SHELF (THE BARRIER)

Amundsen returns 26 January

Oates sacrifices himself
17 March

Beardm●
Glacier
83.5-85°

## SCOTT

| Cape Evans | 80° South | 81° South | 82° South | 83° South | The Gateway |
|---|---|---|---|---|---|
| 1 November | 18 November | 22 November | 27 November | 1 December | 4 December |

One Ton Depot

Scotts final camp, 29 March

ROSS ICE SHELF (THE BARRIER)

Evans dies
17 February

# 60. *The race that never was*

**THE STORY OF AMUNDSEN AND SCOTT'S** famous expeditions to the South Pole has been told and re-told many times over the last century. Their journeys are often framed as a race, but they were not. Scott had planned a slow siege-like attempt to reach 90° South; he had a gentleman's agreement that his would be the only expedition to attempt the Pole that season. He was not aware of Amundsen's mission until their ships sighted each other just before overwintering. Amundsen had changed his plan at the last minute. His original intent had been to reach the North Pole. By swapping to a journey southwards he was risking everything – his money, his livelihood, his reputation. Even the Norwegian government

*(Continued overleaf)*

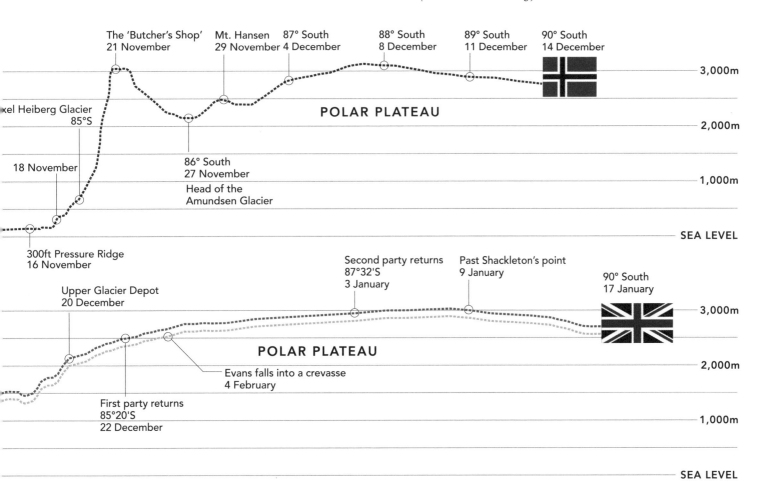

had advised him not to go, but the explorer knew that, if he could beat Scott and become the first person to the Pole, any previous agreement would soon be forgotten in the tide of glory that such an achievement would carry.

Amundsen planned to go fast and light. One of his greatest advantages was his efficient and ruthless use of dogs (he took twice as many as he needed so he could use the spare ones as food for the others), but his route was also critical. He correctly surmised that setting off from the Bay of Whales, on the eastern side of the Ross Sea, would mean that the trip to the Pole would be shorter. It also, and possibly more critically, meant that the Transantarctic Mountains were further south and the traverse at altitude across the high, cold polar plateau would be much shorter. To do this he had to find a new route through the mountains. This was his greatest gamble. If he could not find a passage through the hills and onto the plateau beyond, his prospects were over. As winter turned to spring he was eager to go, but his first attempt in September was beaten back by bad weather. When he eventually started his second trip, however, he still set off a full two weeks before Scott.

Scott, on the other hand, was not set up for speed; he had never planned for a race. His trip was as much a scientific expedition as one of exploration.

He intended to ascend the Beardmore Glacier, a route that Shackleton had used before him. It would be slow, but he knew it would get him up onto the polar plateau. His reluctance to use dogs would be a critical factor, but other things, such as the weather and accidents along the way, would also prove pivotal in the tragedy that unfolded.

I have always wondered about the physiological effect of reaching the Pole only to find Amundsen's flag already there. Would they have made it back if they had got there first? We will never know, but what we do know is what happened on their return trip. Slowed by Evans' fall on the descent of the Beardmore Glacier and Oates' frostbite, they battled towards the coast. Plagued by storms and unseasonably cold temperatures, they ran out of supplies and, without the food and fuel to keep them warm, the three remaining companions died in their tent just 11 miles from the supply depot that would have been their salvation.

*'We shall stick it out to the end, but we are getting weaker, of course, and the end cannot be far. It seems a pity, but I do not think I can write more. R. SCOTT. …For God's sake look after our people.'*

Captain Scott's last entry in his diary, 29 March 1912

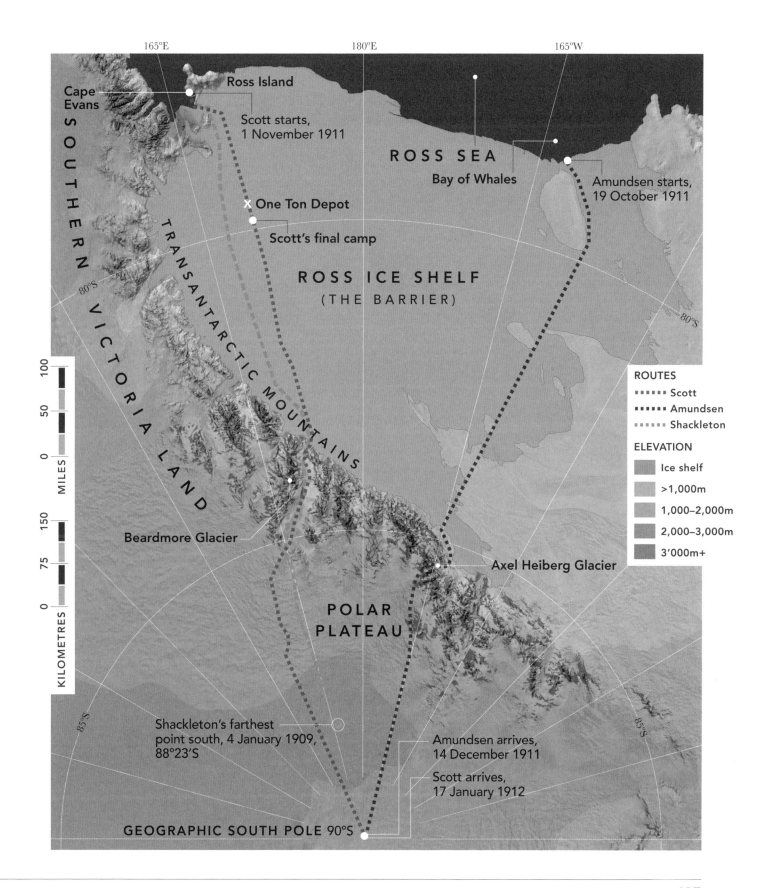

165°E      180°E      165°W

Cape
Evans

Ross Island

Scott starts,
1 November 1911

ROSS SEA

Bay of Whales

Amundsen starts,
19 October 1911

S
O
U
T
H
E
R
N

V
I
C
T
O
R
I
A

L
A
N
D

TRANSANTARCTIC MOUNTAINS

One Ton Depot

Scott's final camp

ROSS ICE SHELF
(THE BARRIER)

80°S

80°S

100

50

0

MILES

150

75

0

KILOMETRES

Beardmore Glacier

Axel Heiberg Glacier

POLAR
PLATEAU

ROUTES
······ Scott
······ Amundsen
······ Shackleton

ELEVATION
Ice shelf
>1,000m
1,000–2,000m
2,000–3,000m
3'000m+

85°S

85°S

Shackleton's farthest
point south, 4 January 1909,
88°23'S

Amundsen arrives,
14 December 1911

Scott arrives,
17 January 1912

GEOGRAPHIC SOUTH POLE 90°S

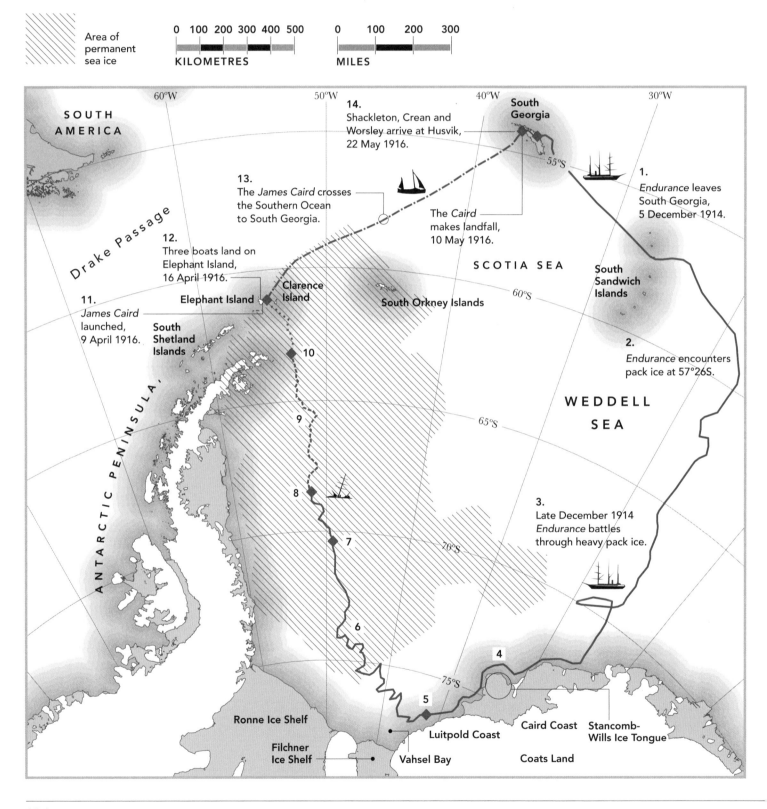

Area of permanent sea ice

0  100  200  300  400  500
KILOMETRES

0    100    200    300
MILES

60°W

SOUTH AMERICA

Drake Passage

50°W

**14.**
Shackleton, Crean and Worsley arrive at Husvik, 22 May 1916.

**13.**
The *James Caird* crosses the Southern Ocean to South Georgia.

**12.**
Three boats land on Elephant Island, 16 April 1916.

**11.**
*James Caird* launched, 9 April 1916.

Elephant Island

Clarence Island

South Shetland Islands

ANTARCTIC PENINSULA

40°W  South Georgia

The *Caird* makes landfall, 10 May 1916.

55°S

SCOTIA SEA

South Orkney Islands

60°S

30°W

**1.**
*Endurance* leaves South Georgia, 5 December 1914.

South Sandwich Islands

**2.**
*Endurance* encounters pack ice at 57°26S.

WEDDELL SEA

10

9

65°S

8

7

70°S

6

5

4

75°S

Ronne Ice Shelf

Filchner Ice Shelf

Vahsel Bay

Luitpold Coast

Caird Coast

Coats Land

**3.**
Late December 1914 *Endurance* battles through heavy pack ice.

Stancomb-Wills Ice Tongue

# 61. *Get on your knees and pray*

**AFTER AMUNDSEN AND SCOTT** reached the Pole in 1911–12, only one great quest was left for polar exploration – the crossing of the continent. Ernest Shackleton, already the veteran of two attempts at the Pole, was the first to try, setting off from South Georgia in December 1914. But, as so often happens in Antarctica, the elements had other ideas. Shackleton's plan was to follow a course charted earlier by the Scottish explorer William Spiers Bruce and land a party in Coats Land, south of the Weddell Sea, but the sea ice that year was particularly bad and his ship, the *Endurance*, never even reached the coast. Beset by impenetrable ice floes in January 1915, the ship was caught fast and drifted, trapped in the ice, for more than nine months. Finally, on 21 November, the crippled ship sank, leaving Shackleton and his crew alone on the ice, with the ship's lifeboats, equipment and rations. They tried pulling the boats over the ice, but the rough terrain proved too difficult and, as the ice was continually drifting northwards, they decided to sit tight and let the currents transport them to the open ocean. On 9 April 1916 the pack ice started to break up and the party took to the three small lifeboats. They rowed across the stormy Bransfield Strait, eventually landing on

### KEY TO MAP NUMBERS

4. The ship navigates along the coast looking for a place to land.
5. *Endurance* trapped in ice, 18 January 1915.
6. Winter 1915, *Endurance* drifts trapped in the sea ice.
7. *Endurance* crushed, ship abandoned, 27 October 1915.
8. *Endurance* sinks, 21 November 1915.
9. Crew drift northwards on the sea ice.
10. Sea ice breaks up, crew take to boats, 9 April 1916.

the uninhabited Elephant Island. The conditions on the island were appalling. With no hope of rescue and winter fast approaching, Shackleton decided to risk a boat trip across the Scotia Sea to South Georgia. It was 1,200 kilometres in an open rowing boat over the roughest sea in the world. If they sailed past the small island in the vastness of the Southern Ocean, Shackleton and his men would have been doomed.

They adapted their lifeboat, the *James Caird*, giving it a makeshift cover and mast. Six men, including Shackleton and Frank Worsley, the ship's captain, were chosen to make the trip. They set off on 24 April, and their sixteen-day boat journey has become the stuff of sailing legend. Soaked through, freezing cold and battered by hurricane-force winds, navigating by starlight and dead-reckoning, they barely made it to South Georgia.

They landed on the south coast of the island. The only habitation they knew of – the whaling station of Stromness – was on the north coast. To reach it they would have to cross the island, a feat that no human had yet accomplished. In order to do this they had to ascend unclimbed ridges and ice caps with only the rags that clothed them. To provide grips as they climbed, they pushed screws salvaged from their boat into their rotting shoes.

Leaving three companions at the boat, Worsley and Tom Crean accompanied Shackleton up the first slope. Initially, the going was slow, through deep snow up the steep ice field. By the first night

*(Continued overleaf)*

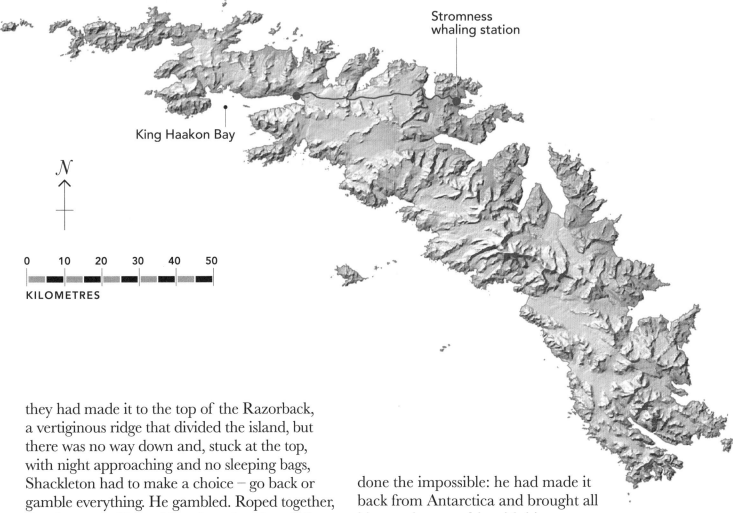

Stromness
whaling station

King Haakon Bay

*N*

0   10   20   30   40   50

KILOMETRES

they had made it to the top of the Razorback, a vertiginous ridge that divided the island, but there was no way down and, stuck at the top, with night approaching and no sleeping bags, Shackleton had to make a choice – go back or gamble everything. He gambled. Roped together, they jumped off the ridge, dropping then sliding on their coiled ropes down to the snowfield 600 metres below. Elated, they carried on through the night and, although they had to retrace their steps several times, they finally made it to the whaling station together, on 22 May 1916.

After several attempts, they managed to rescue the men left on the coast of South Georgia and then those on Elephant Island. Shackleton had done the impossible: he had made it back from Antarctica and brought all his men home safely with him.

*'For scientific discovery, give me Scott; for speed and efficiency of travel, give me Amundsen; but when you are in a hopeless situation, when you are seeing no way out, get down on your knees and pray for Shackleton.'*

Explorer Raymond Priestley's view when asked who was the greatest Antarctic explorer.

## SHACKLETON'S CROSSING OF SOUTH GEORGIA

**1.** Shackleton and his five companions make landfall in Haakon Bay on the south coast of South Georgia. Three of the men – Shackleton, Worsley and Crean – decide to make the perilous crossing of the island, never before attempted, to reach civilization on the north coast.

**2.** Through knee-deep snow the adventurers struggle up the Murray Snowfield.

**3.** After several attempts they reach the top of the Razorback, a steep ridge dividing the island, only to find that there is a sheer precipice on the other side. Rather than turn back they trust to luck and jump – sliding down the snow-slope on their coiled ropes to the level ground below.

**4.** Elated by their escape, they cross the Nineteen Sixteen Glacier by moonlight.

**5.** The party mistakes Fortuna Glacier for Stromness Bay, before realizing their mistake and continuing eastwards.

**6.** They climb and descend Breakwind Ridge into Fortuna Bay. As dawn breaks they faintly hear the whistle of Stromness whaling station in the distance.

**7.** They climb over the ridge of Busen Peninsula and down a narrow gully towards the whaling station.

**8.** The three men limp into Stromness whaling station and are met with disbelief by the whalers there. After two years on the ice Shackleton and his men had found salvation.

*(See pages 170-71 for main map)*

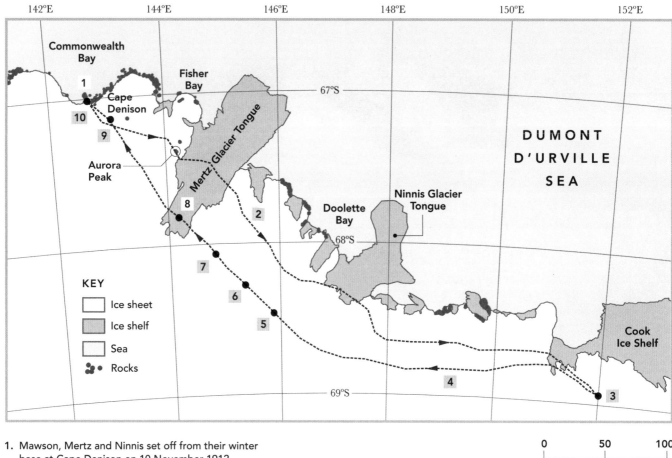

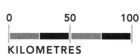

1. Mawson, Mertz and Ninnis set off from their winter base at Cape Denison on 10 November 1912.

2. Initially all goes well and they cross the glaciers later named Mertz and Ninnis making good speed.

3. On 12 December Ninnis is lost down a crevasse.

4. Having lost most of their equipment and dogs, Mawson and Mertz eat their dogs to survive.

5. 28 December they eat their last dog, Ginger. They are forced to 'manhaul' the rest of the way.

6. 30 December, Mertz falls ill.

7. 8 January, Mertz dies.

8. Mawson falls into a crevasse, but manages to pull himself out after several attempts.

9. 19 January, Mawson finds a food depot left by his men.

10. Mawson arrives back on 8 February.

# 62. *The home of the blizzard*

**IMAGINE HOW THE** Australian explorer Douglas Mawson must have felt. Exhausted and at the edge of starvation, he struggled over the last ridge towards his winter base, having lost both his travelling companions on an epic and terrible journey, only to see his ship sailing over the horizon without him.

Luckily, four men had been left behind in the unlikely event of Mawson's return and, although he had to endure a second winter in the Antarctic, Mawson survived. It didn't help that the place they had to inhabit was the windiest spot on Earth. Commonwealth Bay, in East Antarctica, holds the world record for the highest mean annual wind speed – 50mph – with winter winds regularly exceeding 150mph: not a great location to spend a cold, dark Antarctic winter.

Paradoxically, up until Mawson's final journey, his expedition had been one of the most successful in the whole Heroic Age of Exploration. His men had charted the coast of much of East Antarctica, overwintered in two places and surveyed far inland, including making the first journey to the South Geomagnetic Pole. But it is for the last, fateful trip that Mawson will be remembered. His three-man party – himself, Xavier Mertz and Belgrave Ninnis – had set out to chart the Antarctic interior to the east of Commonwealth Bay. Initially, all had gone well. It was at the farthest point, as they were about to return, that disaster struck. One of the sledges, the one carrying the tent and almost all the food, disappeared down a huge crevasse, taking Ninnis with it. He was never seen again.

Mawson and Mertz were left with no tent and little food, hundreds of kilometres from their base on the coast. They started to head back, but the food soon ran out and they had to resort to eating their dogs, one by one, the weakest first, until there were no dogs left and they were forced to haul the sledge themselves. They both became seriously ill, poisoned by eating the livers of the dogs, which contain dangerously high levels of vitamin K, a substance deadly to humans. Mertz could not continue, so Mawson dragged him on the sledge, covering only a few kilometres each day. Mertz deteriorated and, on 8 January, he passed away in the makeshift tent they had constructed as a shelter. Now Mawson was on his own. He stumbled onwards, pulling the sledge. It was that sledge that saved him, for he, too, was suddenly engulfed, falling through a fragile snow bridge into an unseen crevasse. The sledge slid towards the hole but stuck at the top, too large to fit down the crack. Mawson was too exhausted to haul himself out, and it took several hours and a superhuman effort before he scrambled back over the lip.

Willpower alone seemed to drive him the last part of the journey. As he got closer to home, he found food depots left by his companions for his return trip. He got back weeks late, his ship gone, his clothes, health and wellbeing in tatters. But he made it back, in one of the truly heroic exploits of polar exploration.

*(Continued overleaf)*

## THE JOURNEY OF THE *AURORA*: MAWSON'S EXPLORATION OF EAST ANTARCTICA

0    100    200    300    400    500
**KILOMETRES**

Blue shading denotes areas of heavy pack ice

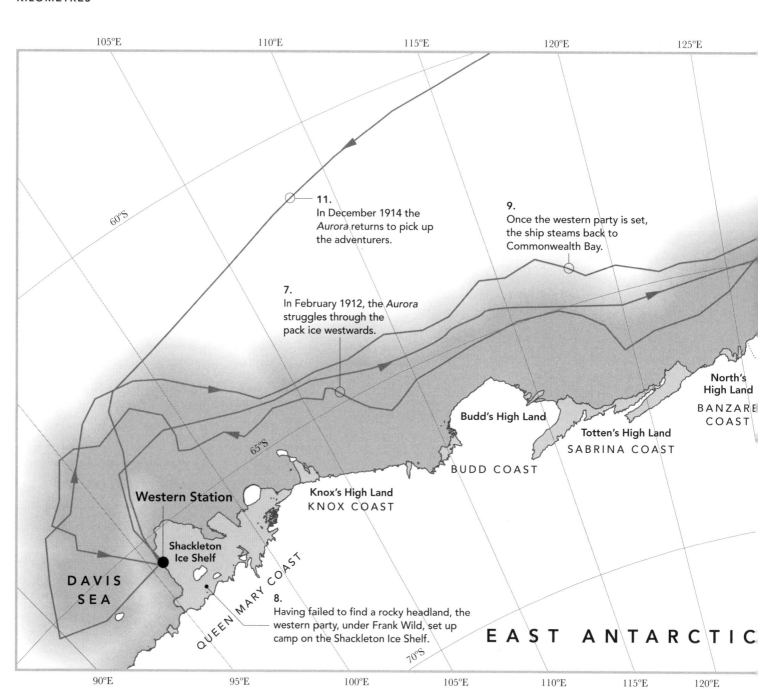

**11.**
In December 1914 the *Aurora* returns to pick up the adventurers.

**9.**
Once the western party is set, the ship steams back to Commonwealth Bay.

**7.**
In February 1912, the *Aurora* struggles through the pack ice westwards.

North's High Land

BANZARE COAST

60°S

Budd's High Land

Totten's High Land

SABRINA COAST

65°S

BUDD COAST

**Western Station**

Knox's High Land

KNOX COAST

Shackleton Ice Shelf

DAVIS SEA

QUEEN MARY COAST

**8.**
Having failed to find a rocky headland, the western party, under Frank Wild, set up camp on the Shackleton Ice Shelf.

EAST ANTARCTIC

70°S

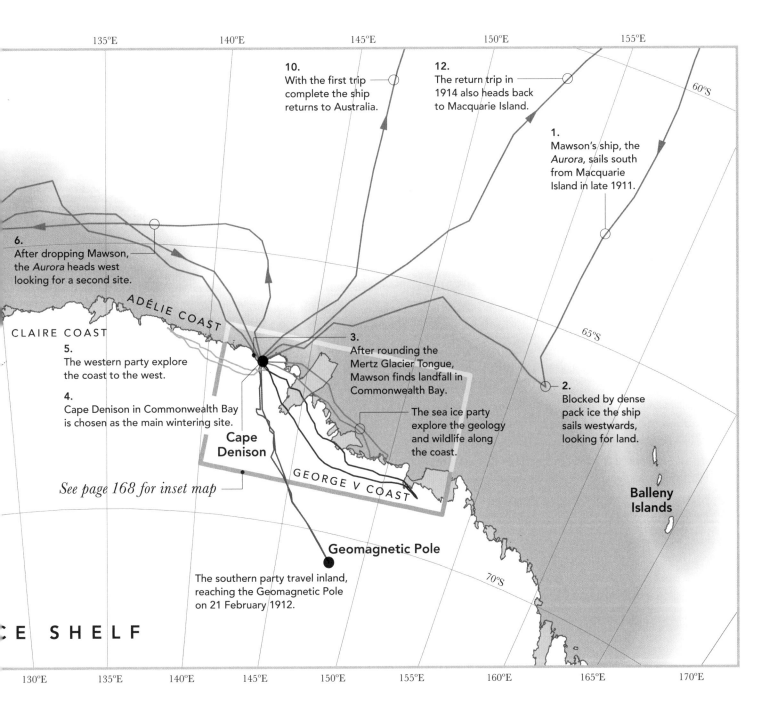

135°E  140°E  145°E  150°E  155°E

60°S

**10.**
With the first trip
complete the ship
returns to Australia.

**12.**
The return trip in
1914 also heads back
to Macquarie Island.

**1.**
Mawson's ship, the
*Aurora*, sails south
from Macquarie
Island in late 1911.

**6.**
After dropping Mawson,
the *Aurora* heads west
looking for a second site.

ADÉLIE COAST

CLAIRE COAST

**5.**
The western party explore
the coast to the west.

**3.**
After rounding the
Mertz Glacier Tongue,
Mawson finds landfall in
Commonwealth Bay.

65°S

**2.**
Blocked by dense
pack ice the ship
sails westwards,
looking for land.

**4.**
Cape Denison in Commonwealth Bay
is chosen as the main wintering site.

The sea ice party
explore the geology
and wildlife along
the coast.

**Cape
Denison**

*See page 168 for inset map*

GEORGE V COAST

**Balleny
Islands**

**Geomagnetic Pole**

The southern party travel inland,
reaching the Geomagnetic Pole
on 21 February 1912.

70°S

CE SHELF

130°E  135°E  140°E  145°E  150°E  155°E  160°E  165°E  170°E

# 63. *Exploring from above*

**IT WAS TEN YEARS** after the end of the First World War that Antarctic exploration truly restarted. In 1928, Hubert Wilkins, an Australian explorer, made the first flight in Antarctica. Wilkins flew over the length of the Antarctic Peninsula and his exploits showed the potential of aerial surveying, paving the way for many others. Aeroplanes would enable explorers and surveyors to chart the harsh continental interior, more quickly, safely and efficiently than on the ground. Within a year, US navy lieutenant Richard E. Byrd had flown to the South Pole and back, and in 1935 the American Lincoln Ellsworth crossed the continent from the Peninsula to the Ross Sea by plane. The more traditional modes of sledge and ship were still utilized, and much of the remaining unknown coastline was charted during this period. Sledging expeditions remained important. One example was the British Graham Land Expedition led by an Australian, John Rymill, who explored the southern section of the Antarctic Peninsula and finally proved that it was part of the main land mass of Antarctica and not an island, as some explorers had previously hypothesized.

However, as international political tensions intensified in the late 1930s governments began to play an increasingly dominant role in Antarctic exploration. They financed major missions such as the German Antarctic Expedition of 1938, an aerial survey that allegedly dropped thousands of iron swastikas on to the ice of Dronning Maud Land to claim the region for Germany. In 1940, Byrd's third US expedition to Antarctica was financed by government agencies. It seemed that the era of the gentleman explorer was over – and that Antarctica was becoming a pawn in a global power struggle.

**EXPLORERS** (SHIP)

— Wilkins (1928–30)
— Larsen (1928–9)
— Byrd (1928–30)
— Riiser-Larsen (1929–30)
— Mawson (1929)
— Isachsen (1930–31)
— Christensen (1938–9)

**EXPLORERS** (AIRCRAFT)

······ Wilkins (1928–9)
- - - Byrd (1928–30)
-·-·· Ellsworth (1935)
······ Ellsworth (1939)
······ Rymil (1934–7)
······ Ritscher (1938–9)

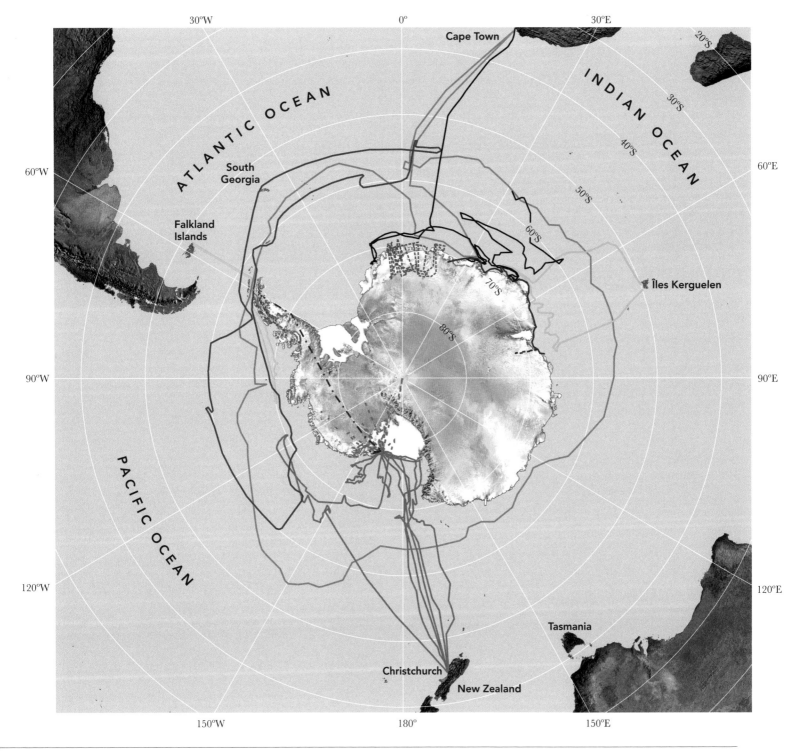

## POSTWAR POWER PLAY
(1944–58)

EXPEDITIONS (AIRCRAFT) ✈ (TRACTOR) 🚜

— US Navy operations
High Jump and Deep Freeze
— Byrd
— Ronne
···· USSR

— Norway / UK / Sweden
— France
— Australia
— Commonwealth Trans-Antarctic Expedition
■ Base

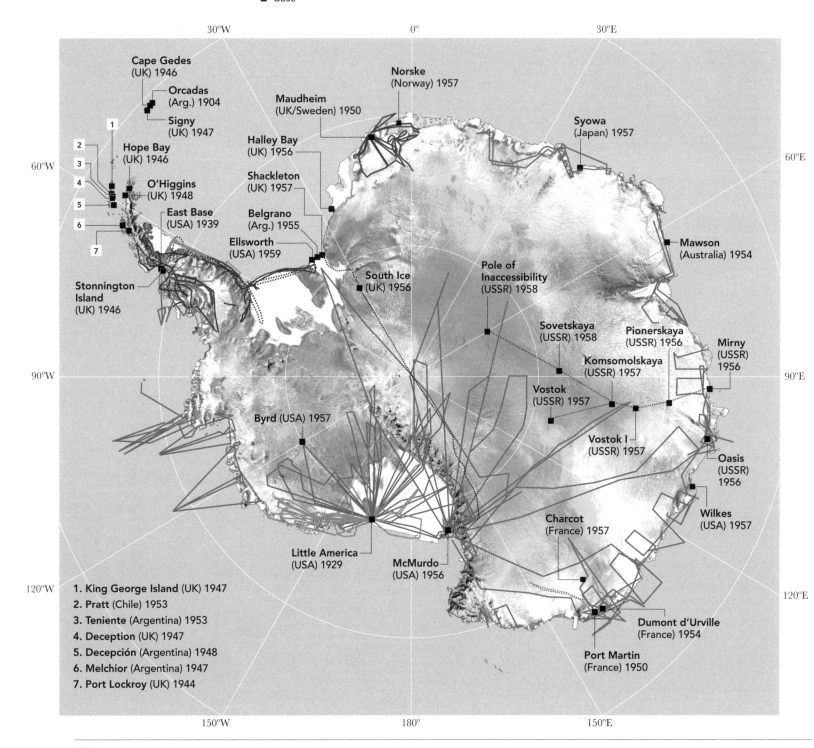

Cape Gedes (UK) 1946
Orcadas (Arg.) 1904
Signy (UK) 1947
Hope Bay (UK) 1946
O'Higgins (UK) 1948
East Base (USA) 1939
Stonnington Island (UK) 1946

Norske (Norway) 1957
Maudheim (UK/Sweden) 1950
Halley Bay (UK) 1956
Shackleton (UK) 1957
Belgrano (Arg.) 1955
Ellsworth (USA) 1959
South Ice (UK) 1956

Syowa (Japan) 1957
Mawson (Australia) 1954
Pole of Inaccessibility (USSR) 1958
Sovetskaya (USSR) 1958
Pionerskaya (USSR) 1956
Mirny (USSR) 1956
Komsomolskaya (USSR) 1957
Vostok (USSR) 1957
Vostok I (USSR) 1957
Oasis (USSR) 1956
Wilkes (USA) 1957
Charcot (France) 1957
Dumont d'Urville (France) 1954
Port Martin (France) 1950

Byrd (USA) 1957
Little America (USA) 1929
McMurdo (USA) 1956

1. King George Island (UK) 1947
2. Pratt (Chile) 1953
3. Teniente (Argentina) 1953
4. Deception (UK) 1947
5. Decepción (Argentina) 1948
6. Melchior (Argentina) 1947
7. Port Lockroy (UK) 1944

# 64. Postwar power play

AFTER THE SECOND WORLD WAR, America took the lead in exploring unknown parts of the continent. Richard E. Byrd, already a veteran of Arctic and Antarctic exploration from the interwar period, was given command of a large flotilla of ships and planes in the US-led Operation High Jump, which was to become the largest Antarctic exploration to-date. In 1946, they set up their main base on the Ross Ice Shelf, at a place they named Little America, but they also flew sorties from the aircraft carrier *USS Philippian Sea*. Fourteen ships, six helicopters, six flying boats, around twenty small aircraft and 4,700 men were at Byrd's command for this truly momentous expedition. The planes systematically surveyed the regions around the Ross Sea and the Transantarctic Mountains, making many discoveries. From the aircraft carriers they were able to map large portions of the coast and hundreds of miles inland. High Jump was followed in the 1950s by Operation Deep Freeze, another US military expedition, also led by Byrd. From 1956 onwards, a number of permanent US bases were built under Deep Freeze, including McMurdo Station, Bay of Whales (near Little America) and the South Pole Station (later renamed Amundsen–Scott Station). This operation expanded the American exploration outwards, further into East and West Antarctica.

Other Americans were also active. Finn Ronne, a Norwegian-born US citizen, privately financed and led the 1946–7 Ronne Antarctic Research Expedition, which charted and explored the southern part of the Antarctic Peninsula and Weddell Sea both by plane and dog sled.

Meanwhile, the British were active in the Peninsula region. The secret wartime Operation Tabarin, initiated in 1944, was tasked with setting up almost a dozen permanent bases. After the war, these stations became part of the Falkland Islands Dependency Survey, a UK-government operation to give precedence to the UK's territorial claim on the area. To cement their claim, the British surveyed and mapped much of the region in the 1950s, first by dog sled and then in a concerted aerial-survey programme. Other countries were also using the presence of feet-on-the-ground military personnel to justify their Antarctic claims. These included Argentina and Chile, whose claims in the Peninsula region overlapped with Britain's. Each set up a number of military bases in the 1940s and 1950s. Tensions were growing, and at one stage shots were exchanged between Chilean and Argentine troops on the South Shetland Islands.

The other postwar superpower, the USSR, also started to become involved. The Russians had a long history of Antarctic exploration, from Bellingshausen onwards. In the mid 1950s they set up bases around Prydz Bay in East Antarctica and, using tractor trains, travelled far inland to establish stations at the Geomagnetic Pole and the Pole of Inaccessibility.

Antarctica was increasingly becoming a theatre of operations in which governments would flex their military muscle and show dominance. Mapping and exploration had become a political tool in the quest for Antarctic sovereignty.

# 65. *The scientific age*

**WITH THE SIGNING** of the Antarctic Treaty in 1959 (see pages 132–33), Antarctica entered a new age. Military bases were replaced with scientific stations and an era of international collaboration ensued. The focus of exploration also changed. By the end of the International Geophysical Year in 1957–8, most of the rocky areas of Antarctica had been surveyed, although it would take a decade or more to convert the thousands of aerial photos into useful maps. What was left was the ice, the huge white wilderness that covered over 99 per cent of the continent. Many parts of the icy landscape had already been flown over, but it was not what was at the surface that interested the scientists; what fascinated them was what was under it. Aerial survey in the 1950s had led to the invention of the radio-echo sounder, an instrument that could measure the thickness of the ice, and this, along with seismic surveys, gravimeters and magnetometers, gives information about the make-up of the rocks beneath kilometres of ice. These new scientific methods were employed to make subglacial maps and investigate the hidden landscape.

Huge geophysical aerial-survey campaigns set up through various national Antarctic programmes or larger academic institutions started to collect data on ice thickness and the geology beneath the vast ice sheets. Many discoveries were made, and many questions answered. How much ice was there? What did the subglacial landscape look like? Was the continent beneath the ice the same shape as that above? How had Antarctica formed? Many of the maps on the preceding pages (see Draining the Depths, page 40; and Hidden World, page 50) have been built on data collected by these surveys.

Gaps still remain in the geophysical map of Antarctica, but these hidden areas – possibly the last totally unknown places on Earth – are systematically being surveyed, and soon the continent will be fully mapped, both on the surface and under it.

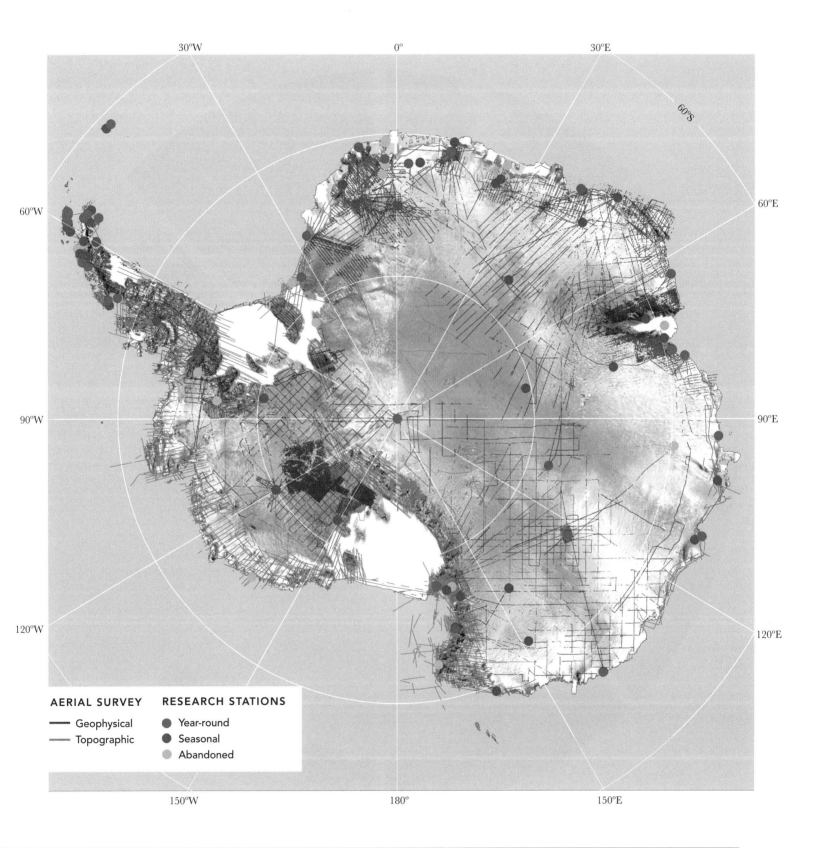

AERIAL SURVEY
— Geophysical
— Topographic

RESEARCH STATIONS
● Year-round
● Seasonal
● Abandoned

SURFACE ROUGHNESS

VELOCITY

FALSE COLOUR

OPTICAL

SEA-ICE CONCENTRATION

SURFACE TEMPERATURE

ELEVATION

SNOW THICKNESS

# 66. *The satellite age*

**SINCE THE FIRST SATELLITES** were launched into space in the 1950s and 1960s, they have played an important part in Antarctic mapping. Over the last two decades, this influence has grown, and today most surveying on the continent is done by satellite.

The number and range of observation satellites that are taking images and measurements of the Earth's surface are astounding. Almost everything that we want to chart and measure can now be achieved from space. Of course, for Antarctica, where the harsh environmental conditions make work on the ground very difficult, being able to study things from the comfort and warmth of your own office is a great benefit. It is also cheaper and safer. Whether you are counting penguins, monitoring vegetation, mapping the geology or the flow of the glaciers, the range of what you can achieve using satellite data is incredible.

The map here shows some of the applications of satellite imagery. Different satellites can do different things, so let's go through a few of them.

- *Velocity* By looking at change over time, satellites can measure how fast things move. This is especially useful for measuring ice-sheet flow. When a glacier loses ice it usually speeds up, so keeping an eye on the velocity can be an early warning system to tell us about glaciers that are changing.

- *Optical* Normal optical satellite images are like a camera. They use light visible to the naked eye and give an excellent impression of the ice surface. This easily understood output is regularly used for background mapping. The resolution of these images tends to be higher than those from other types of satellite.

- *Surface temperature* By monitoring thermal radiation, satellites can track changes in temperature, which is useful for studying the ice sheets and sea ice.

- *Snow thickness* Satellites with microwave instruments can measure how far their pulses of energy penetrate the snow pack to reveal snow thickness and recent snow accumulation.

- *Elevation* A variety of sensors, including optical, radar and laser instruments, have been used to construct elevation models of the continent and to monitor the changing height of the ice sheet.

- *Sea-ice concentration* Radar satellites penetrate through the clouds to give an uninterrupted view of sea-ice conditions – a vital tool for ship navigation in the Southern Ocean.

- *False colour* By recording parts of the electromagnetic spectrum invisible to the human eye, such as infrared light, satellites can identify a plethora of different objects. Different satellites record different wavelengths of light, breaking the spectrum up into different bands. Our eyes measure three types of light – red, blue and green – but many satellites measure many more. Some, the 'hyperspectral satellites', measure hundreds of discrete parts of the spectrum. All sorts of things can be mapped using this imagery: vegetation, rocks or penguin guano, among others.

- *Surface roughness* Radar instruments record surface roughness. This data is used in a variety of ways, including detecting change and giving information about the type of snow surface and the amount of meltwater.

# 67. A most historic place

**OF ALL THE PLACES** in Antarctica, Ross Island must rank as the most historic.

This icebound island in the southern portion of the Ross Sea is dominated by the presence of the permanently active volcano Mount Erebus. It has seen many expeditions, scientists and adventurers over the last two centuries and has borne witness to tragedy and triumph. Today, it hosts the largest human settlement in Antarctica – McMurdo Station, run by the US scientific programme. In the past, several of the continent's most famous explorers used the island as a base on their

expeditions to the South Pole, and in many cases their huts and artefacts remain, preserved by the cold, dry polar air. But the island has also played host to the greatest loss of life, when, in 1979, a sightseeing airliner crashed into Mount Erebus at Lewis Bay, killing all 237 people on board.

This map shows the historic sites on the island, as well as the two research stations and the locations of penguin colonies and peaks. Also listed are some of the main characters that have contributed to the story of the island.

*Captain* James C. Ross
*1800–1862*

One of Antarctica's most celebrated mariners. Between 1839 and 1842 he explored and charted both the Ross and Weddell seas and discovered, in 1841, the island which now bears his name. He also named the prominent volcanic peaks of Erebus and Terror after his two ships.

*Captain* Robert F. Scott
*1868–1812*

A pivotal figure in the island's history. He visited the island twice and used it as the main base for his famous expeditions.

He also named the island for Ross, and he and his party climbed and explored its peaks, building a number of huts on its shoreline.

Edward Wilson
*1872–1912*

Participating in both Scott's expeditions, Scott's second in command was a celebrated naturalist and artist. He led the 'Worst Journey in the World' to Cape Crozier to recover a penguin egg in the depths of winter.

Sir Ernest Shackleton
*1874–1922*

Was part of Scott's first expedition. Later he returned to the island, using it as a base for his own attempt at the Pole.

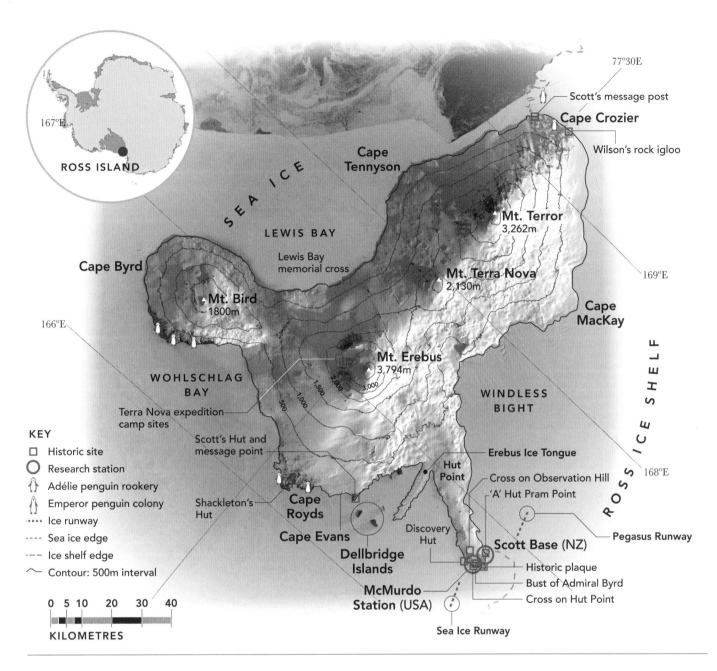

ROSS ISLAND

167°E

77°30E

Scott's message post

Cape Crozier

Wilson's rock igloo

169°E

Cape
Tennyson

SEA ICE

LEWIS BAY

Mt. Terror
3,262m

Cape Byrd

Lewis Bay
memorial cross

Mt. Terra Nova
2,130m

Cape
MacKay

166°E

Mt. Bird
1800m

WOHLSCHLAG
BAY

Mt. Erebus
3,794m

1,500
2,000
3,000
500
1,000

WINDLESS
BIGHT

R O S S   I C E   S H E L F

168°E

Terra Nova expedition
camp sites

Scott's Hut and
message point

Erebus Ice Tongue

Hut
Point

Cross on Observation Hill

'A' Hut Pram Point

Pegasus Runway

Shackleton's
Hut

Cape
Royds

Discovery
Hut

Scott Base (NZ)

Cape Evans

Dellbridge
Islands

Historic plaque

Bust of Admiral Byrd

Cross on Hut Point

McMurdo
Station (USA)

Sea Ice Runway

**KEY**

▫ Historic site
○ Research station
🐧 Adélie penguin rookery
🐧 Emperor penguin colony
···· Ice runway
---- Sea ice edge
-·- Ice shelf edge
⌒ Contour: 500m interval

0  5 10   20   30   40

**KILOMETRES**

*Rear Admiral*
Richard E. Byrd
*1888–1957*

Byrd had a long and
distinguished career of polar
exploration at both poles. He
pioneered the use of aircraft
in Antarctica, led the huge
American postwar efforts to
map the continent, and was
the first person to fly over the
South Pole. He also set up
McMurdo Station.

Sir Edmund Hillary
*1919–2008*

After Hillary's Everest
fame, he set his sight on
Antarctica. As part of the
Trans-Antarctic Expedition
in 1958 he became only the
third person to reach the
Pole and the first by using a
motorized vehicle. He was
instrumental in setting up
Scott Base for New Zealand.

**181**

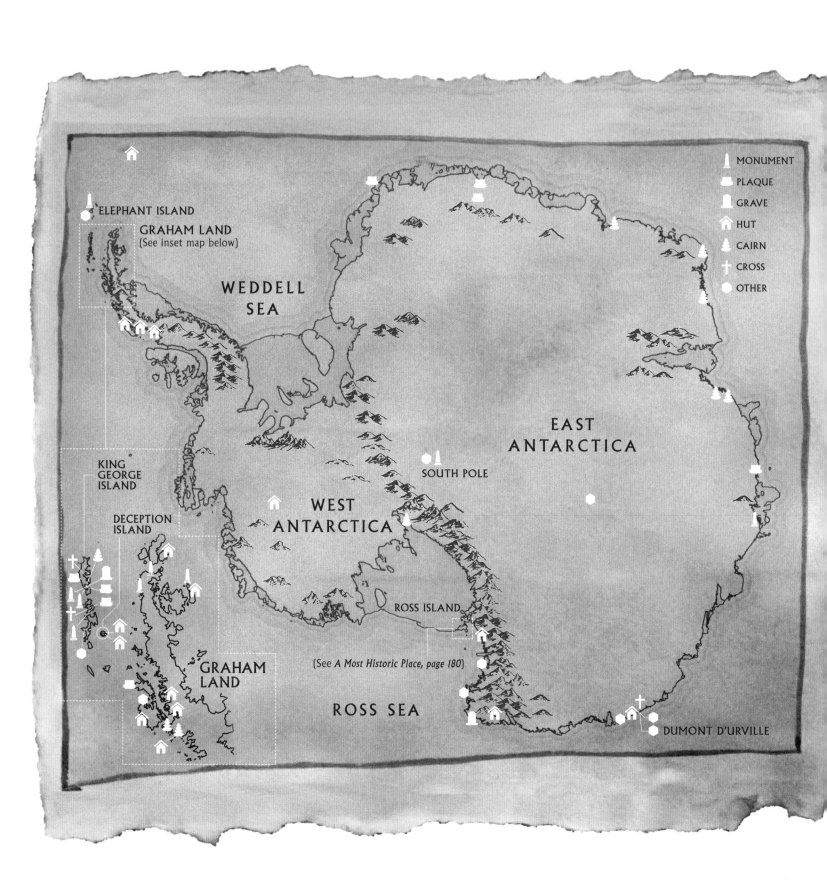

MONUMENT

PLAQUE

GRAVE

HUT

CAIRN

CROSS

OTHER

ELEPHANT ISLAND

GRAHAM LAND
(See inset map below)

WEDDELL
SEA

EAST
ANTARCTICA

KING
GEORGE
ISLAND

SOUTH POLE

DECEPTION
ISLAND

WEST
ANTARCTICA

ROSS ISLAND

GRAHAM
LAND

[See *A Most Historic Place*, page 180]

ROSS SEA

DUMONT D'URVILLE

# 68. *Traces of the past*

**HUMAN HISTORY ON** the Antarctic continent may only be two hundred years old, but those two centuries have left a plethora of historic sites and monuments. Many of the explorers from the Heroic Age of Exploration and quite a few from more recent times are commemorated at various sites around the continent in memorials, plaques and statues. Tragedies in the pursuit of knowledge on this most harsh of continents are also remembered in this way. In addition to these memorials, there are the remains of ancient huts and research bases, in various states of repair, across Antarctica. Some, like Scott's Hut on Ross Island, are almost perfectly preserved and look as if the explorers left only yesterday, while in other huts and buildings only the foundations or crumbling ruins remain. Eighty-five of these sites are currently protected as official Historic Sites and Monuments under the Antarctic Treaty System. This number fluctuates regularly as nations propose (or occasionally de-list) memorials.

This map shows the geographic spread of these sites, displayed by type. I have grouped the various locations into six classes: memorials (which contains statues and busts), plaques, huts, cairns, crosses and a miscellaneous 'other' category. This category includes such diverse things as lighthouses, shipwrecks, historic post-boxes, tractors and ice caves, which do not fall easily into the other groupings. It is not surprising that the geographic spread of these sites is centred around the location of research stations, as people, and the history which they become, are heavily focused on these places.

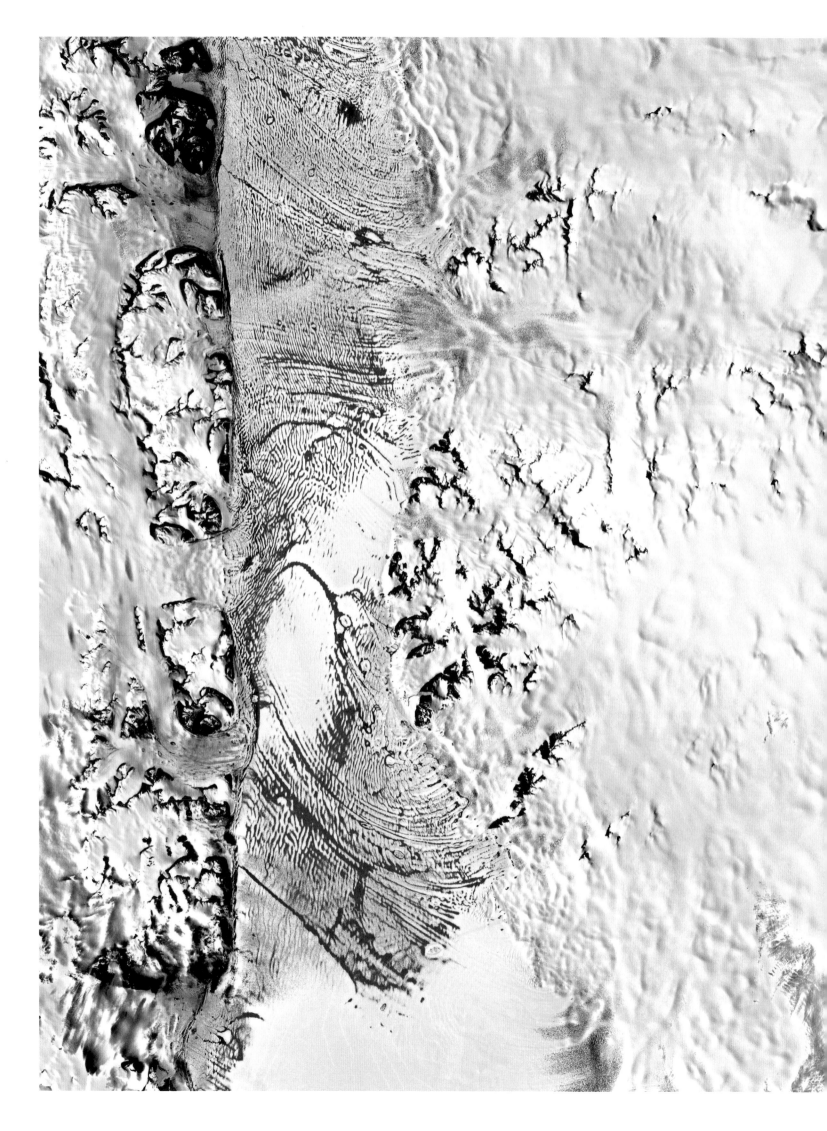

# CHAPTER 9

# *Future*

69. Looking ahead *186*

70. Antarctica 10,000: The distant future *188*

# 69. *Looking ahead*

**SCIENTISTS PREDICT THAT**, if the world continues to warm, Antarctica will start to melt. Modelling exactly how long this melt will take and which parts will melt first is a complex task. Many environmental processes have to be understood before this can be done effectively. The map uses the most up-to-date models available to show a comparison of how the continent looks today and what it will look like in the future. In this example, we use the year 2500 to compare with the present.

The map shows quite a lot of change, especially in West Antarctica. Here, over 70 per cent of the ice has been lost, creating a completely new sea, tentatively called the West Antarctic Sea. Some of the larger mountains near the coast remain as islands. One of the other most significant changes is the demise of the great Antarctic ice shelves; for example, the Ross Ice Shelf is a fraction of its former self and the Ronne Ice Shelf disappears completely. The Antarctic Peninsula will become an island, separated from the rest of the continent by a channel running through the area where the Evans Ice Stream now resides, although what the new island's name will be is difficult to predict.

In East Antarctica, several coasts will shrink inwards, creating large new bays around Wilkes and George V Land. Overall, the size of the continent will shrink by around a fifth, to just over 11 million square kilometres, but in the centre of the land mass the extreme height and cold at the heart of the ice sheet will protect it from any sign of warming. Here, the ice surface will remain remarkably unchanged. In this scenario, around 10 per cent of Antarctica's ice is lost. This, along with other melting glaciers around the world, would raise global sea levels by around 10 metres, flooding many coastal towns and cities such as London, New York and Tokyo. Indeed, eight of the world's top ten most populous cities would be inundated by the sea if this were to happen.

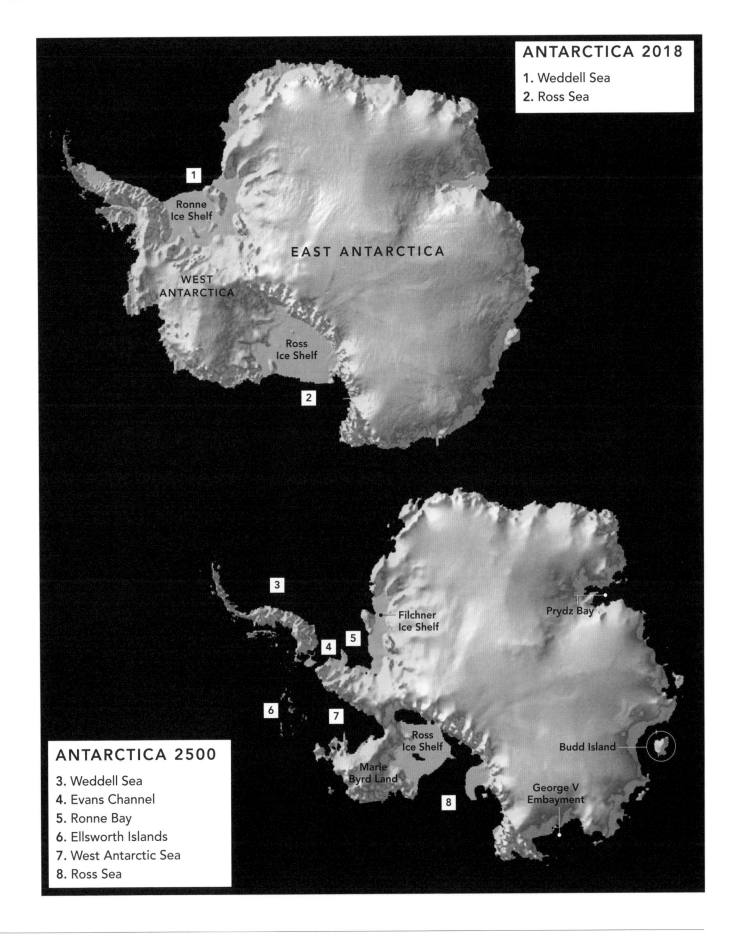

**ANTARCTICA 2018**

1. Weddell Sea
2. Ross Sea

EAST ANTARCTICA

Ronne
Ice Shelf

WEST
ANTARCTICA

Ross
Ice Shelf

**ANTARCTICA 2500**

3. Weddell Sea
4. Evans Channel
5. Ronne Bay
6. Ellsworth Islands
7. West Antarctic Sea
8. Ross Sea

Filchner
Ice Shelf

Prydz Bay

Ross
Ice Shelf

Marie
Byrd Land

Budd Island

George V
Embayment

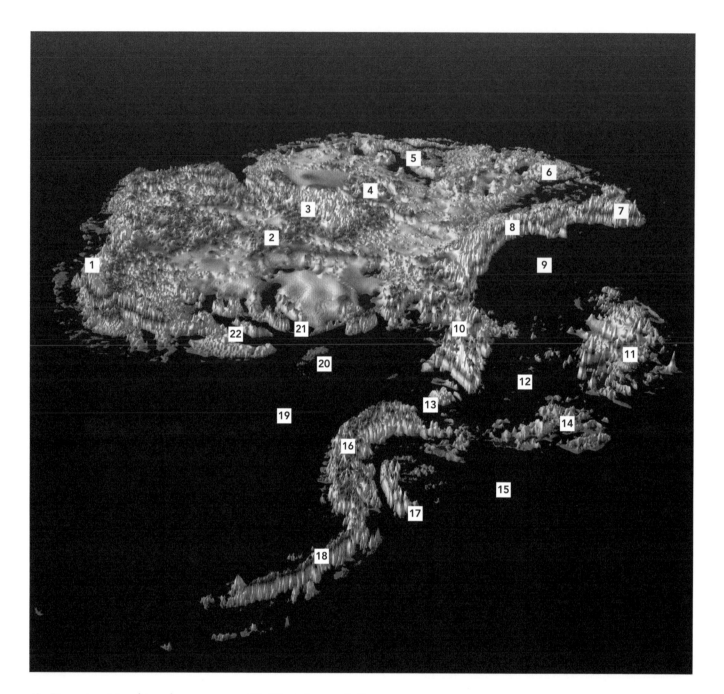

1. Dronning Maud Land
2. Lambert Uplands
3. Gamburtzev Mountains
4. Vostok Uplands
5. Aurora Sea
6. Wilkes Sea
7. Victoria Land
8. Transantarctic Mountains
9. Ross Sea
10. Hollick-Kenyon Peninsula
11. Marie Byrd Land
12. West Antarctic Sea
13. Evans Strait
14. Ellsworth Land
15. Bellingshausen Sea
16. Palmer Land
17. Alexander Island
18. Graham Land
19. Weddell Sea
20. Berkener Island
21. Recovery Fjord
22. Bailey Fjord

# 70. Antarctica 10,000: The distant future

**SCIENTISTS HAVE CALCULATED** that, if humanity exploits all the known reserves of coal, oil and gas on the planet, the greenhouse gases that will be released will eventually melt all the ice in Antarctica. Whether this will ever happen, and how long it will take, is a matter of speculation. The vast amounts of ice covering the continent are so large they would take thousands of years to thaw. We know that at the end of the last glacial period the huge ice caps over Europe and North America took around ten thousand years to melt, so here we have used that timescale as a benchmark of what could happen.

If Antarctica did ever melt completely, the continent would be very different. The visualization here gives an oblique perspective of what this land mass might look like.

Removing the ice reveals that the underlying bedrock is not a single island but an archipelago of smaller islands surrounding the larger mass of East Antarctica. Several other things also happen when you remove the icy cover from the continent. Firstly, the sea level rises. Assuming that Greenland and other glaciers also melt, seas globally would rise by around 66 metres, so the continental land mass, and everywhere else on the planet, would effectively be 66 metres lower. However, in Antarctica, this rise is dwarfed by what happens to the land itself. The ice sitting on Antarctica weighs a lot; on average, each square kilometre of the continent has almost 2,000 tonnes of ice resting above it. This burden presses down on the Earth's crust under the ice sheet, depressing it below its normal level. If the ice melted, the crust would 'rebound', lifting the land up to a kilometre where the ice is thickest. This would mean that almost 4 million square kilometres of land – an area equivalent to all the countries of the European Union – would rise out of the sea.

This map takes these processes into account to give a hypothetical view of an Antarctica far in the future. The names on the map have been derived from existing place names.

Glossary *193*

Acknowledgements *197*

References, data sources and further reading *199*

Index *203*

Photographs *207*

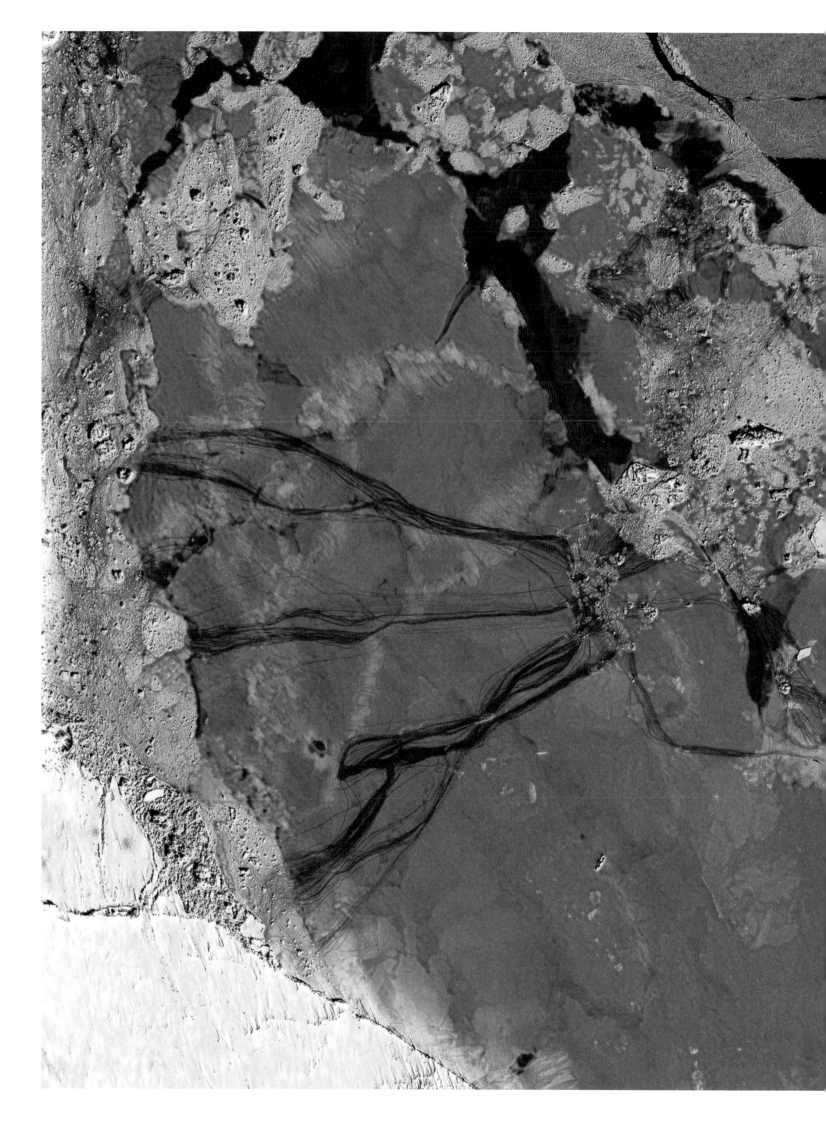

# *Glossary*

*Arid*

An environment that has little or no rain; too dry or barren to support vegetation.

*Atmospheric column*

A way of describing the different features found in the Earth's atmosphere at different heights.

*Basin*

A circular or oval valley or natural depression on the Earth's surface, especially one containing water.

*Bergy bit*

A small-to-medium piece of floating ice formed when a piece of ice breaks off and lands in the water. Generally showing between 1 and 5 metres above sea level, with a length of between 5 and 15 metres.

*Biomass*

The total quantity or weight of organisms in a given area.

*Blocky iceberg*

A flat-topped iceberg with steep vertical sides.

*Boulder pavement*

A flat surface of boulders where the finer matrix of particles between them has been removed to leave just a stony pavement.

*Calving*

The splitting off of a mass of ice from an iceberg or glacier, often with a loud, explosive sound.

*Centrifugal force*

A force moving in a circular path and directed away from the centre around which an object is moving.

*Craton*

A large, stable block of the Earth's crust forming the nucleus of a continent.

*Crustacean*

An invertebrate animal with a hard shell and several pairs of legs, e.g. a crab or a woodlouse.

*Dome iceberg*

A round-topped iceberg.

*Drydock iceberg*

An iceberg that has been eroded in the middle to form a U-shaped enclosure.

*Eddy*

A circular movement of water causing a small whirlpool.

*Embayment*

A recess in a coastline forming a bay.

*Endemism*

Plants and animals that are both native of a particular area and restricted to it.

*Equinox*

The time or date when the sun crosses the celestial equator, when day and night are of equal length (approx. 22 September and 20 March).

*Fast ice*

Ice which covers the seawater and is attached to land.

*Firn*

Crystalline or granular snow, especially on the upper part of a glacier, where it has not yet been compressed into ice.

*Food web*

A system of interlocking and interdependent food chains.

*Frazil ice*

Tiny, needle-like ice crystals 3 to 4 millimetres in diameter, suspended in the top layer of the ocean water, which represent the first stages of sea-ice growth.

*Geographic South Pole*

The bottom of the world; the axis on which the Earth spins.

*Geologist*

An expert in the study of rocks.

*Geophysicist*

Someone who studies the Earth using gravity, magnetic, electrical and seismic methods.

*Glacial ice*

Thick layers of snow are gradually compressed into glacial ice. In Antarctica, it can take hundreds or even thousands of years for enough snow to accumulate to form ice.

*Gondwana*

An ancient supercontinent that broke up about 180 million years ago. It split into land masses we recognize today.

*Grease ice*

Ice particles which are clumped together; visually akin to an oil slick on water.

*Growler iceberg*

Small chunks of floating ice that rise only about 1 metre out of the water.

*Gyre*

A spiral or vortex.

*Hydrothermal vent*

A volcanic fissure in the seabed out of which heated, mineral-rich water flows.

*Hyper-aridity*

Used of an area of extreme dryness.

*Hyper-saline*

Used of harsh environments that have salt concentrations much greater than that of seawater. Only organisms with extreme adaptations can tolerate these conditions.

*Ice core*

A cylinder of ice drilled out of an ice sheet or glacier. The deepest can be up to 3 kilometres long.

*Ice shelf*

Floating glacial ice, often hundreds of metres thick, that has flowed from the continent out to sea (but is still attached to the land).

*Land ice*

Ice sheets on land, including mountain glaciers and ice sheets covering Greenland and Antarctica.

*Lead*

A channel of water in heavy sea ice.

*Moratorium*

A temporary prohibition of an activity.

*Névé*

Partially compacted granular snow that forms the surface of the upper end of a glacier; broadly, a field of granular snow.

*Nunatak*

The exposed rocky part of a mountain sticking out of an ice sheet.

*Ozone*

An isotope of oxygen ($O_3$) commonly found as a gas.

*Pack ice*

Broken pieces of sea ice (frozen sea), which can be heavy (with few leads, or loose) with open water between the pieces of frozen sea.

*Pancake ice*

A floating circular piece of ice between 30 centimetres and 3 metres in diameter. They can be up to 10 centimetres thick, with raised edges caused by compression when they bash together.

*Pangea*

In early geologic time, a supercontinent formed 335 million years ago that incorporated almost all the land masses on Earth.

*Pelt*

An animal skin, usually of fur.

*Phytoplankton*

Tiny floating plants, often single-celled organisms, which photosynthesize in the oceans.

*Pinnacle iceberg*

An iceberg with one or more spires.

*Polar front*

An area where cold polar water meets warmer subtropical water. Here there is a convergence zone where cold Antarctic water sinks under warmer sub-Antarctic water.

*Polar vortex*

An upper-level, low-pressure large area lying near the Earth's poles which rotates counter-clockwise at the North Pole (a cyclone), and clockwise at the South Pole.

*Polynya*

A stretch of open water surrounded by ice.

*Radio-echo sounding*

A technique used by glaciologists to measure the internal structure, ice thickness, size and shape of ice masses by using radiowaves.

*Saline*

Used of a salty liquid.

*Sea mount*

A submerged mountain.

*Sea stack*

A column of steep rock rising out of the sea, remaining after the erosion of cliffs.

*Seabed plateau*

An area of flat ocean floor that often rises 2 to 3 kilometres above the surrounding sea floor with steep, sloping sides.

*Seismology*

The analysis of vibrations in the Earth's crust or in solid ice.

*Sonar*

A technique using sound waves to measure distance in water, in this context to measure water depths beneath a vessel.

*Spreading ridge*

A fracture zone along the ocean bottom where molten mantle material comes to the surface, thus creating new crust.

*Stewardship*

The job of supervising or taking care of something, such as an organization or property.

*Stratosphere*

The second major layer of the Earth's atmosphere, just above the troposphere and below the mesosphere. It begins around a height of 7 kilometres, near the Poles.

*Stratovolcano*

A volcano, typically conical in shape, built up of alternate layers of lava and ash.

*Subducting*

A process in plate tectonics where the edge of one crustal plate descends below the edge of another.

*Subduction zone*

Where one tectonic plate descends beneath another.

*Subglacial*

Used of an area situated underneath a glacier or ice sheet.

*Subglacial trough*

A channel or trough that has formed beneath a glacier or ice sheet.

*Tabular iceberg*

A steep-sided iceberg with a flat top, usually formed by the calving of an ice shelf. These are the largest types of iceberg.

*Tectonic plate*

A large, moving piece of rock that forms the Earth's crust. Tectonic plates are constantly moving and give rise to mountains, earthquakes and volcanoes when they collide.

*Thermoline circulation*

A massive current of water moving around the globe, from northern to southern oceans and back again. Warm surface waters sink down, and cold, nutrient-rich waters push upwards.

*Transantarctic Mountains*

The mountains that stretch across Antarctica, dividing the continent into east and west. The highest peak in the range is Mount Kirkpatrick, at 4,528 metres.

*Transverse fault*

A fault that occurs when a block of rock fractures and two blocks of rock slide past each other in opposite directions. The motion is predominantly horizontal.

*Turbo-prop aeroplane*

An aeroplane with a turbojet engine used to drive an external propeller.

*Vortex*

A circling mass of water or air.

*Water column*

Used of the different features found in seawater at different depths.

*Wedge iceberg*

An iceberg with a top that narrows to a pyramidal shape.

*White ice*

Coarse, granular, porous ice (as in a glacier) that is usually formed by compaction of snow and appears white.

# *Acknowledgements*

**I AM INCREDIBLY FORTUNATE** to work in a place with some of the best scientists and most knowledgeable people on polar issues, history and politics. Many of them have helped in data gathering, research and proofing of this book, and I would like to thank them all. This work would not have been possible without them.

I would especially like to thank my wife, Lisa, for her patience, keen eye and helpful suggestions, not to mention the excellent artistry of many of the accompanying drawings in this book. Also, a special thanks to several of my colleagues at British Antarctic Survey (BAS) who proofread various chapters; David Vaughan for the Ice, Mike Meredith for the Sea, Adrian Fox for the Exploration chapter, and Kevin Hughes for his advice on the wording of several of the more politically sensitive maps in the People section.

Many scientists helped with individual maps, several of them from the British Antarctic Survey (those named here without their institution are all from BAS).

Gareth Rees, from the Scott Polar Research Institute, with whom I have worked for many years, calculated the data for The Day the Night and the Half-light map. Teal Riley and Alex Burton-Johnson advised on the Geological Setting map. Additional data from the Geomap of Antarctica project was also given to me from Alan Aitken by the University of Western Australia. The Making of Antarctic map would not have been possible without the help of Tom Jordan, who taught me Geoplates software and advised on the map itself. John Smellie from the University of Leicester and Rob Larter both helped and advised on the Antarctic Volcanoes, and Rob was extremely helpful in the making of the Quaking Sea map, both giving advice and calculating the data for the earthquakes. Kevin Hughes and Pete Convey advised in the Alien Invasion map.

The Atmosphere chapter was one on which my knowledge was quite thin, so I am much indebted to John King about what to map, who to talk to and for his ideas on the Polar Vortex graphic. Tom Bracegirdle gave me his new data on temperature projects for Tricky Predictions, and Steve Colwell and Paul Breen helped me gather data for The Hole at the Bottom of the World. The text for his map, about the ozone hole, was checked and added to by Jon Shanklin, one of the scientists who discovered the ozone hole in 1984.

Mike Meredith helped immensely with the Sea chapter and was very supportive. He advised on the Ocean Currents map and suggested chatting to Dave Munday, who produced unpublished data from his oceanographic model to map complexity in the ocean, which was then worked into the Ocean Energy map. Thomas Frölicher, from Princeton University, gave data from his recent publication for the Earth's Lungs visualization.

The Wildlife chapter would not have been possible without the help of several colleagues with whom I have worked closely over the years; Phil Trathan and Jen Jackson for data from the IWC that was used in The Blood-red Sea piece, Iain Staniland for advice on the International Seal Travels work and Richard Phillips and Henri Weimerskirche for the albatross tracks in The Great Wanderers map.

Jamie Oliver, our graphics artist, was kind enough to let me use and expand his excellent graphic of the Penguin Line-up – I hope that the final version lives up to your original, Jamie.

On the People chapter, John Eager, Neil Cobbet and Richard Warren offered essential advice on the Halley maps in the Moving Home Antarctic Style section. Susie Grant provided GIS data and Mark Belchier gave advice for the Exploiting the Ocean Map, while Heather Lynch from Stony Brook University in New York gave data for the Lure of Antarctica.

For the last chapter I am indebted to Rob DeConto, from the University of Massachusetts for providing exciting new data on future ice-retreat to map the Into the Future graphic.

I also would like to thank all my family, friends and colleagues, especially from the BAS Mapping and Geographic Information Centre, for their support and for putting up with my endless babble and often inane enthusiasm for the project.

Thank you also to Maria Bedford and to Tony Lyons for his fantastic graphic work.

Finally, I would like to say a huge thank you to Penguin Random House and especially Cecilia Stein; this is her project as much as mine and would never have happened without Cecilia's vision, enthusiasm and hard work. Cecilia, I hope that the book lives up to your initial idea.

# References, data sources and further reading

**THERE ARE MANY** excellent reference books about Antarctica. I have read and digested many of them over the past twenty years, far too many to catalogue here. Below is a list of some of the more specific sources of data used to formulate the maps in this book, as well as some possible further reading that you may enjoy if you have been inspired by our maps.

## General data sources

The coastlines and topographic data of Antarctica for the maps has been taken from the Antarctic Digital Database: https://www.add.scar.org/

Place names are taken from the Antarctic Placename Committee website: https://apc.antarctica.ac.uk/ and the Composite Gazetteer of Antarctica: https://www.scar.org/data-products/place-names/ and https://data.aad.gov.au/aadc/gaz/scar/

Other data sources include: *The Times Comprehensive Atlas of the World* (14th edn), Open Streetmap, maps by the British Antarctic Survey, such as BAS Miscellaneous Series Sheets 15A and 15B Antarctica and the Arctic and BAS Miscellaneous Series Sheets 13A and 13B Antarctic Peninsula and the Weddell Sea and Graham Land and the South Shetland Islands. Another widely used layer utilized for almost all maps of the seabed was the General Bathymetric Chart of the Oceans (2014) and the selection of USGS 1:200,000 and other maps available from the Polar Geospatial Center: https://www.pgc.umn.edu/data/maps/

## Specific data sources

**Ice thickness** Data taken from Bedmap2: P. T. Fretwell, H. D. Pritchard et al. (2013), *Cryosphere*, 7, 375–93, 2013; https://doi.org/10.5194/tc-7-375-2013.

**Velocity** Data taken from The Measures project: E. Rignot, J. Mouginot, B. Scheuchl (2011), *Science*, 333 (6048); doi: 10.1126/science.1208336.

**A changing world** Data for sea-ice change taken from S. Stammerjohn, R. Massom, D. Rind and D. Martinson (2012), 'Regions of rapid sea ice change: An inter-hemispheric seasonal comparison', *Geophysical Research Letters*, 39 (6); https://doi.org/10.1029/2012GL050874.

**Ice-shelf-elevation change** Data taken from M. McMillan, A. Shepherd et al. (2014), 'Increased ice losses from Antarctica detected by CryoSat-2', *Geophysical Research Letters*, 41 (11); https://doi.org/10.1002/2014GL060111.

**Drowning coasts** Data used to construct seamless high-resolution regional elevation models was Aster GDEM available from the USGS Earth Explorer portal: https://earthexplorer.usgs.gov/; the global elevation model was the ETOPO1 dataset: https://www.ngdc.noaa.gov/mgg/global/global.html.

**The anatomy of an ice sheet** Data taken from Bedmap2: P. T. Fretwell and H. D. Pritchard et al. (2013), *Cryosphere*, 7, 375–3, 2013; https://doi.org/10.5194/tc-7-375-2013

**Draining the depths** Map interpreted from Bedmap2 and S. J. Clark et al. (2013), 'Potential subglacial lake locations and meltwater drainage pathways beneath the Antarctic and Greenland ice sheets', *Cryosphere*, 7, doi: 10.5194/tc-7-1721-2013.

**Antarctic time machine** Information taken from various sources online.

**Shrinking shelves** Data taken from the Antarctic Digital database and A. J. Cook and D. V. Vaughan (2010), 'Overview of areal changes of the ice shelves on the Antarctic Peninsula over the past 50 years', *Cryosphere*, 4; www.the-cryosphere.net/4/77/2010/.

**Geological setting** Map interpreted from G. E. Grikrov and G. Leychenkov, Tectonic map of Antarctica (2012), CCGM-CGMW, with the help of Alex Burton-Johnson and Teal Riley of British Antarctic Survey

**The hidden world** Information from a merging of M. Morlighem (2019) MEaSUREs BedMachine Antarctica, Version 1, http//doi.org/10.5067/C2GFER6PTOS4, and modelled data currently under review: P. T. Fretwell et al. (submitted), 'Improved ice sheet bed topography from satellite images'.

**The making of Antarctica** Constructed using Geoplates software and data with the help of Tom Jordan from British Antarctic Survey.

**Southern Ocean volcanoes** Information taken from W. E. LeMasurier et al. (1990) *Volcanoes of the Antarctic Plate*

*and Southern Oceans*, Washington, DC, American Geophysical Union, 487 pp.; https://doi.org/10.1029/AR048.

**The quaking sea** Earthquake data derived from http://www.isc.ac.uk/ehbbulletin/ and E. R Engdahl, et al. (1998), 'Global teleseismic earthquake relocation with improved travel times and procedures for depth determination', *Bulletin of the Seismological Society of America* 88, 72–43.Coasts and bathymetry from GEBCO 2014. Geological information derived from various sources, including BAS Tectonic map of the Scotia Arc, updated with advice from R. Larter.

**The driest place on earth** Elevation and contours constructed from Aster GDEM data available from the USGS Earth Explorer portal: https://earthexplorer.usgs.gov/. Other topographic information from the Antarctic Digital database. Additional details taken from a range of online sources, including the PGC map Protecting Antarctica's McMurdo Dry Valleys (2011), ANT REF-ES2004-003, available from http://maps.apps.pgc.umn.edu/id/133.

**Alien invasion** Plotted from information gathered from Y. Frenot et al. (2005), 'Biological invasions in the Antarctic: extent, impacts and implications', and S. L. Chown and P. Convey (2016), *Antarctic Entomology: Annual Review of Entomology*, 61, and P. Convey and M. Lebouvier (2009), 'Environmental change and human impacts on terrestrial ecosystems of the Sub-Antarctic islands between their discovery and the mid-twentieth century', *Procedures of the Royal Society of Tasmania*, 143 (1).

**Mountains** Elevation models, hill shading and contours constructed from Aster GDEM available from the USGS Earth Explorer portal: https://earthexplorer.usgs.gov/.

**Don't forget your thermals** Wind vectors and temperature ramp made from the RACMO climate model of Antarctica: https://www.projects.science.uu.nl/iceclimate/models/racmo.php.

**The hole at the bottom of the world** Map data taken from NASA Ozone Watch: https://ozonewatch.gsfc.nasa.gov/ Ozone spiral data taken from BAS internal records.

**A future in our hands** Data kindly provided by T. Bracegirdle from the Antclim21 projects; see https://www.scar.org/science/antclim21/data-surface-projections/.

**Stormy weather** Images taken from the NASA EOSDIS WorldView portal: https://worldview.earthdata.nasa.gov/.

**Polar vortex** Elevation models used to construct the map were taken from the Bedmap2 database.

**The Southern Ocean** Bathymetry data taken from GEBCO 2014: https://www.gebco.net/data_and_products/gridded_bathymetry_data/.

**Islands in the stream** Imagery is Landsat8 taken from the Earth Explorer website: https://earthexplorer.usgs.gov/; coastlines are from the Antarctic Digital Database.

**Ocean currents** The line of currents have been drawn from Y. S. Kim and A. H. Orsi (2014), 'On the variability of Antarctic Circumpolar Current fronts inferred from 1992–2011 Altimetry', *Journal of Physical Oceanography*, 44; https://doi.org/10.1175/JPO-D-13-0217.1.

**Ocean eddies** Map constructed from a new data-driven model by David Munday of British Antarctic Survey.

**The greatest change on Earth** Sea ice data from

**The engine of the ocean** Schematic map drawn from J. Marshall and K. Speer (2011), 'Closure of the meridional overturning circulation through Southern Ocean upwelling', with additional help from Mike Meredith.

**The life of a berg** Iceberg tracks downloaded from the Antarctic Iceberg Tracking database: http://www.scp.byu.edu/data/iceberg/ by J. Budge and D. G. Long.

**The green ocean** Chlorophyll data taken from the NASA MODIS portal using chlorophyll-a concentration data at: https://modis.gsfc.nasa.gov/data/dataprod/chlor_a.php. Information on the phytoplankton was from *Antarctic Marine Protists* by F. J. Scott and H. J. Marchant (eds.), 2005. Illustrations drawn by Peter and Lisa Fretwell.

**Earth's lungs** Data from T. L. Fröchler et al. (2014), 'Dominance of the Southern Ocean in anthropogenic carbon and heat uptake in CMIP5 Models', *Journal of Climate*, 28; https://doi.org/10.1175/JCLI-D-14-00117.1.

**Keystone krill** Krill data from the KrillBase database, with kind permission of Angus Atkinson. The food web infographic was constructed from various sources. Hand drawn illustrations by Lisa Fretwell.

**The realm of the emperor** Information to construct this map was taken from the supplemental material of S. Jenouvrier et al. (2017), 'Influence of dispersal processes on the global dynamics of Emperor penguin, a species threatened by climate change', *Biological Conservation*, 212.

**An ocean of penguins** Distribution information taken from the IUCN AND Birdlife international datazone: http://datazone.birdlife.org/home.

**International seal travels** The marine mammal data were collected and made freely available by the International MEOP Consortium and the national programmes that contribute to it (http://www.meop.net), using the QGIS3 database.

**The blood-red sea** Whale catch data are from the IWC database.

**The great wanderers** Albatross tracks kindly provided by H. Wiemerskirch and R. Phillips. Hand drawn illustration by Lisa Fretwell.

**The richest place on earth** Information has be derived from the South Georgia GIS; https://www.bas.ac.uk/project/sg-gis/; additional locations of king penguin colonies are from personal information given by Sally Poncet, Phil Trathan and Norman Ratcliffe.

**Going south** Plotted using information from the COMNAP website: https://www.comnap.aq/SitePages/Home.aspx; and from various BAS and internet sources.

**Who lives there?** Figures used to construct this infographic were taken from the COMNAP Antarctic Station Catalogue 2017: https://www.comnap.aq/Members/Shared%20Documents/COMNAP_Antarctic_Station_Catalogue.pdf.

**Sweet home Antarctica** Information taken from the COMNAP Antarctic Station Catalogue 2017: https://www.comnap.aq/Members/Shared%20Documents/COMNAP_Antarctic_Station_Catalogue.pdf. Hand drawn illustrations by Lisa Fretwell.

**Who owns it?** The lines of the political claims are constructed from BAS internal data.

**Mac Town** Elevation model from the Polar Geospatial Center by A. G. Fountain et al., 'High-resolution elevation mapping of the McMurdo Dry Valleys, Antarctica, and surrounding regions, *Earth System Science Data*, 9, 435–43; https://doi.org/10.5194/essd-9-435-2017, 2017. Names and vector data taken from a combination of Open Streetmap, Google Earth and reference maps from the US Polar Geospatial Center.

**Halley Station** Data taken from internal BAS datasets and base information. Imagery is from the Sentinel2 satellite available from: https://scihub.copernicus.eu/.

**International antics** Data taken from the Antarctic Digital Database and the COMNAP website.

**Antarctic skies** Information gathered from the SCAR Air Operations Planning maps: https://www.scar.org/data-products/air-op-maps/, available through the SCAR map catalogue: https://data.aad.gov.au/aadc/mapcat/list_view.cfm?list_id=57.

**Exploiting the ocean** Data derived from the CCAMLR GIS, provided by David Herbert and Suzie Grant, with helpful advice from Suzie Grant and Mark Belchier. Illustrations by Lisa Fretwell.

**Tourist hub** Information on marine tour ship traffic kindly provided by Heather Lynch from the publication H. J. Lynch et al. (2009), 'Spatial patterns of tour ship traffic in the Antarctic Peninsula region', *Antarctic Science*: doi:10.1017/S0954102009990654. Data and locations of the top tourist sites from IAATO: https://iaato.org/en_GB/tourism-overview#ant_destinations, accessed 2017.

**To find a continent**, **The heroic age**, **Exploring from above** and **Postwar powerplay** The ship tracks, exploration routes and plane flights for the maps have been digitized from a series of A0 maps made by the US Navy-Hydrographic office for the US Antarctic Program in 1958. These rare maps have been very useful, and these and all the other maps in this chapter have been augmented by the information from a number of Antarctic chronicles, including Richard Sale's Polar Reaches and Beau Riffenburgh's Polar Exploration.

**The greatest escape** Maps taken from Otto Nordenskjöld's *Antarctica: Or, Two Years amongst the Ice of the South Pole*.

**The race that never was** Information from various sources; there has been a lot written about Scott's doomed trip; however, the main map was constructed from GDEM and Bedmap2 elevation data. The lines and dates were interpreted from the 1912 Arthur G. Chater map taken from Amundsen's account.

**Get on your knees and pray** Again, no shortage of information here. I have tried to use the original map and text from Shackleton's book *South*. For the South Georgia crossing map, the elevation model was taken from the South Georgia GIS and the route was interpreted from the British Antarctic Survey South Georgia 2018 map.

**The home of the blizzard** Maps interpreted from Mawson's *Home of the Blizzard*.

**The scientific age** Data to draw the lines taken from a range of British Antarctic Survey internal resources.

**A most historic place** Elevation taken from Aster GDEM data; the background image is a Sentinel2 satellite image; other information taken from a variety of sources, including USGS and PGC maps.

**Traces of the past** Information from the Antarctic Digital Database and ATCM: https://documents.ats.aq/ATCM34/WW/atcm34_ww002_e.pdf.

**Looking ahead** Analysis kindly given by Rob DeConto, based upon R. M. DeConto and D. Pollard (2016), 'Contribution of Antarctica to past and future sea-level rise', *Nature*, 531, 591–7; http://dx.doi.org/10.1038/nature17145.

**Antarctica 10,000: The distant future** Constructed by Peter Fretwell from a basic isostatic uplift model (using a constant rheology and a moving window based on ice weighting) using Bedmap2 data with a single sea-level adjustment.

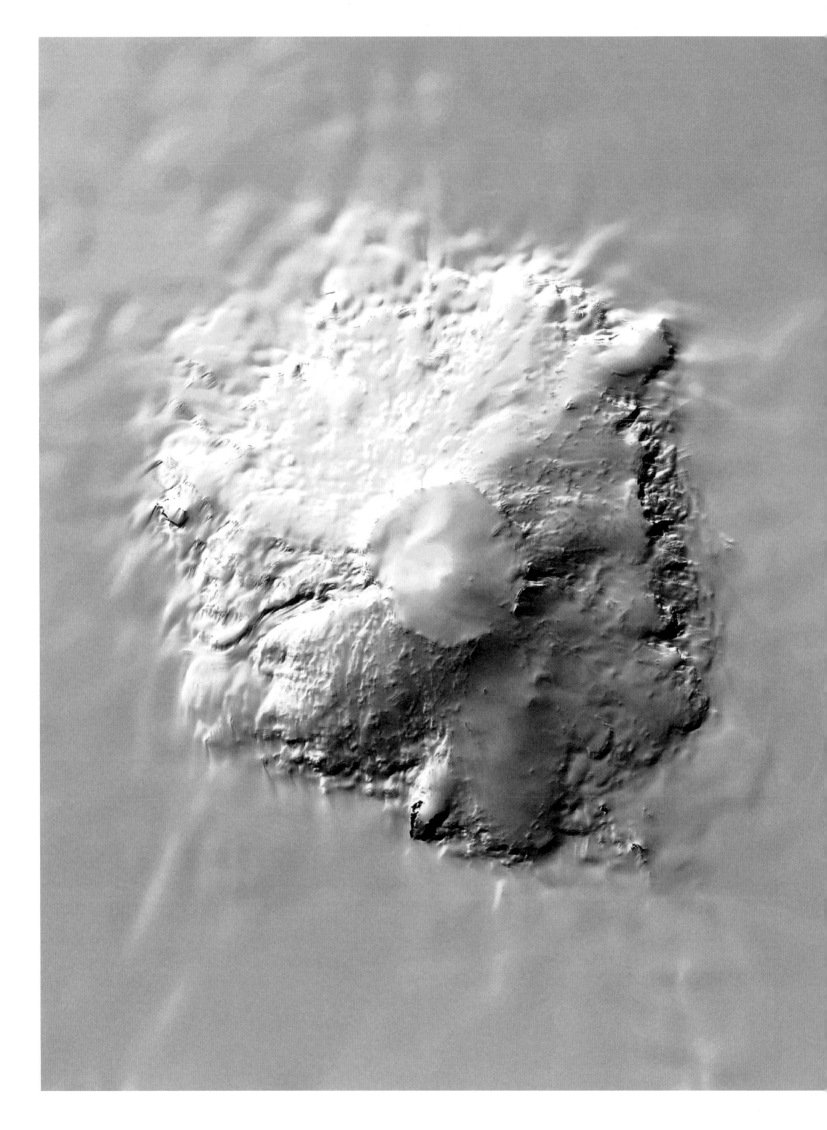

# *Index*

## A

Adélie penguin 113
albatross 80, 88, 119, 120, 121
algae 102
alien species, effect on Antarctica 62, 120
Alfred Wenger Institute 130
Amundsen, Roald 155, 161, 162, 165
Amundsen–Scott Base 127, 128, 130, 175
Amundsen Sea 35
Andes Mountains 14, 52
*Antarctic* (ship) 156, 158
Antarctic Circle 26, 73
Antarctic Circumpolar Current 93, 94, 100
Antarctic Coastal Current 93, 100
Antarctic Peninsula 14, 20, 22, 25, 35, 36, 43, 49, 100, 102, 113, 124, 125, 127, 145, 148, 156, 172, 175, 186
Antarctic Plate 19, 59
Antarctic Sound 156
Antarctic Treaty 19, 127, 132, 135, 176
Argentina 124, 127, 135, 158, 175
Argentina Range 66
Australia 124, 127, 152

## B

Bay of Whales 162, 175
Beardmore Glacier 162
BedMachine Antarctica 50
Belgium 131
Bellingshausen, Fabien von 100, 124, 152

bergs *see* icebergs
Bharati 131
Blood Falls 60
blue whale 116
Borchgrevink, Carsten 155
Boulder Pavement 60
Bransfield, Edward 152
Bransfield Strait 165
Brazil 142
British Antarctic Survey 74
British Graham Land Expedition 172
Bruce, William Spiers 165
Brunt Ice Shelf 130, 138, 141
Bulgaria 142
Byrd, Richard E. 172, 175, 181
Byrd Glacier 30, 32

## C

calderas 55, 57
cats 62
CCAMLR (Commission for the Conservation of Antarctic Marine Living Resources) 146
Canterbury, New Zealand 124, 125
Coats Land 165
Cape Town, South Africa 124, 124
carbon dioxide (CO2) 42, 79, 94, 105
CFCs *see* chlorofluorocarbons
Chile 124, 127, 135, 142, 152, 175
China 130, 142
chinstrap penguin 113
chlorofluorocarbons 74
climate change 36, 42, 45, 79, 83, 105, 110, 120
Cold War 135

Commission for the Conservation of Antarctic Marine Living Resources (CCAMLR) 146
Commonwealth Bay 169
Concordia 128
Cook, James 120, 152
crevasses 38
Crozet Islands 119

## D

De Gerlache, Adrien 155, 148
Deception Island 57, 102
deep water 99
dinosaurs 52
Discovery expedition (Scott) 155
Dobson Units (measure of ozone) 76
Don Juan Pond 60, 142, 145, 152
Drake Passage 93, 124, 131, 172
Dronning Maud Land/Mountains 66
dry valleys *see* McMurdo Dry Valleys
Drygalski Mountains 66

## E

earthquakes 59
East Antarctic ice sheet 17, 25, 30
East Antarctica 14, 17, 35, 36, 49, 52, 69, 169, 175, 186, 189
*Endurance* (ship) 165
ecosystem, Antarctic 62, 109, 114, 121
Ekström Ice Shelf 130
Elephant Island 165, 120
elephant seal 114
Ellsworth, Lincoln 172
Ellsworth Mountains 66, 113

emperor penguin 110
equinoxes 26
erect crested penguin 113
Evans, Edgar 162
Evans Ice Stream 186

**F**

Falkland Islands 59, 124
Falkland Islands Dependency Survey 175
France 127, 128
Fenriskjeften (mountains) 69
Fenristunga (ice cap) 69
Fimbulisen Ice Tongue 132
firn 42
fishing 109, 146
Fiordland penguin 113
fossil fuels 42, 105
fur seals 120, 152
Furious Fifties 80

**G**

gentoo penguin 113
Geographic South Pole 17
geology, of Antarctica 19, 49, 59, 88, 176, 179
Geomagnetic South Pole 17
George V Land 186
Gerlache Strait 148
German Antarctic Expedition 172
Germany 130, 172
glacial ice 25, 35, 36, 42, 105
glaciers 7, 14, 30, 35, 36, 45, 60, 66, 83, 88, 100, 120, 121, 179, 186
global warming *see* climate change
Gondwana 52
Grandidier Channel 148
Graham Land 20, 172
gravity modelling 49, 50
Great Britain 128, 152

greenhouse effect 42, 79, 94
gray whale 116
Greenland 36, 42, 189
grounding line 2

**H**

Halley Research Station 74
Halley VI (*see* also Halley Research Station) 156, 169, 183
'Heroic Age of Exploration' 155
Hillary, Edmund 181
Historic Sites and Monuments (HSMs) 183
Hobart, Tasmania 124
Holtanna (mountain) 69
Hope Bay 156, 158
humpback whale 117
Hut Point Peninsula 56, 137

**I**

ice cores 42
ice sheets 14, 17, 20, 25, 45, 50, 128, 138
ice shelves 14, 19, 25, 32, 92, 100, 114, 138, 141, 152, 156
icebergs 14, 32, 93, 100, 138, 141, 156
'iceberg alley' 100
Imperial Trans-Antarctic expedition 155
India 131
Indian Ocean 119
International Geophysical Year 132, 176
International Whaling Commission 116
Italy 128

**J**

*James Caird* (lifeboat) 165
Jang Bogo (South Korean station) 130
*Jason, The* (ship) 156
Jason Peninsula 156
Joinville Island 156

**K**

Kerguelen 88, 119
King George Island 142
Kinntanna mountain 69
krill 80, 109, 146
Kunlun 130

**L**

Labyrinth, The 60
Lake Vostok 41
Lambert Glacier 30, 32
Larsemann Hills 131
Larsen, Carl Anton 155, 156, 158
Larsen Ice Shelf 45, 156
lava lakes 57
Lemaire Channel 148
Lewis Bay 180
Linnaeus Terrace 60
Little America 175

**M**

Magnetic South Pole 17
Marie Byrd Land 17
Marsh station 145
Mawson, Douglas 73, 155, 169
Maxwell Bay 142
McMurdo Dry Valleys 60, 73
McMurdo Station 137, 175, 181, 180
Mertz, Xavier 169
mice 62, 120, 121
minke whale 117
moss 49
Mount Craddock 66
Mount Erebus 56, 180
Mount Gardiner 66
Mount Kirkpatrick 66
Mount Michael 57
Mount Michael 57
Mount Shinn 66

Mount Siple 55
Mount Takahe 54
Mount Tyree 66
Mount Vinson 66

# N

NASA 60
Neptune's Bellows 57
Neumayer III station 130
Neumayer Channel 148
New Zealand 113, 124, 181
Nimrod expedition 155
Ninnis, Belgrave 169
Nordenskjöld, Otto 155, 156, 158–9
nunataks 50

# O

Oates, Captain Lawrence Captain 162
Operation Deep Freeze 175
Operation High Jump 175
Operation Tabarin 175
Onyx River 60
ozone layer 74, 83

# P

Palmer, Nathaniel 152, 153
Palmer Land 20
Paulet Island 156, 158
penguins 80, 88, 109, 110, 113, 114,
 119, 120, 148
Peru 142
*Philippian Sea*, USS 175
phytoplankton 102, 109
Pine Island Glacier 32, 35
plankton 80
plate tectonics 52, 88
Poland 142
polar front 19, 86, 93, 102, 113
polar vortex 83
Pole of Ignorance 17

Pole of Inaccessibility 17, 175
Pole of Isolation 17
Pole of Mass 17
polynyas 96
Port Stanley, Falkland Islands 124
Priestley, Raymond 167
Prince Gustav Channel 45, 158
Princess Elizabeth Land 17
Princess Elizabeth Station 131
Punta Arenas, Argentina 124

# Q

Queen Elizabeth Land 20

# R

radio-echo sounding 50
rats 62, 120, 121
Razorback, South Georgia 167
Recovery Glacier 30, 32
reindeer 62, 120, 121
right whale 116
Roaring Forties 80
Ronne, Finn 175
Ronne Antarctic Research Expedition 175
Ronne–Filchner Ice Shelf 14, 20, 66, 186
Ross, James C. 180
Ross Ice Shelf 14, 20, 25, 35, 155, 175,
 186
Ross Island 56, 137, 155, 180, 183
Ross Sea 33, 73, 83, 93, 124, 127, 130,
 162, 172
Rothera Research Station 145
royal penguin 113
Russia 127, 142, 152; *see* also Soviet
 Union
Rutford Ice Stream 66
Rymil, John 172

# S

Santiago 152
satellites 179
Saunders Island 57
Scientific Committee on Antarctic
 Research 20
Scotia Arc 59
Scotia Plate 59
Scotia Sea 59, 100, 102, 109, 165
Scott, Robert F. 137, 155, 161, 162, 165,
 180
Scott Base 127, 181
Scott's Hut 18
sea ice 35, 96, 99, 110, 148
sea-ice factories 96
sea-level rise 35, 36
seabirds 88, 114, 119, 120
seals 80, 88, 109, 114, 119, 120, 153
sealers 120, 152, 153
sei whale 116
seismic data 50
Sentinel Mountains 66
Shackleton, Ernest 109, 113, 121, 137,
 155, 162, 165, 167, 180
Smith, William 152, 153
Snow Hill Island 156, 158
soil 49
sonar 50
South Africa 124
South America 14, 52, 124, 124, 142
South American Plate 59
South Atlantic 100
South Georgia 59, 62, 88, 100, 114, 120,
 121, 165, 167
South Korea 130
South Orkney Islands 19, 59
South Pole 17, 19, 26, 127, 130, 155,
 161, 172, 179, 180
South Sandwich Islands 57
South Sandwich Plate 59
South Sandwich Trench 59
South Shetland Islands 59, 142, 148,
 152, 175

South Shetland Plate 59

Southern Ocean 19, 52, 55, 62, 80, 86, 88, 93, 94, 99, 102, 105, 109, 113, 114, 116, 119, 121, 124, 145, 146, 152, 165

Soviet Union 135, 175

sperm whale 116

Stromness 167

Sub-Antarctic Islands 102

subglacial drainage 41

Sweden 156, 158

tabular bergs 38, 100

## T

Taishan station 130

Tasmania 124

Terra Nova Bay 130

thermohaline circulation 99

Thwaites Glacier 32, 35

Tom Crean 167

toothfish 146

Totten Glacier 30, 32

tourism 125, 148

Transantarctic Mountains 14, 20, 65, 155, 162, 175

Trinity Peninsula 152

## U

US Antarctic Program 137, 180

USA 127, 130, 135, 137, 175

Ulvetanna mountain 66, 69

*Uruguay* (ship) 158

Ushuaia, Argentina 124

## V

Victoria Valley 60

Visson Massif 66

volcanoes 55, 54, 88

Vostok Research Station 733

## W

wandering albatross 119

Weddel Sea 93, 100, 156, 165, 175, 180

Weddell seal 114

West Antarctic ice sheet 30

West Antarctica 14, 25, 32, 35, 36, 49, 55, 186

whales/whaling 80, 109, 114, 116, 117, 120, 121

Wilkes Land 186

Wilkins, Hubert 172

Wilson, Edward 180

Worsley, Frank 165

Wright Valley 60

## XYZ

Zongshan station 130

Zucchelli station 130

# Images

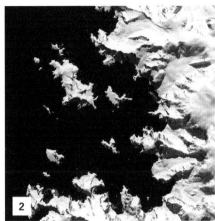

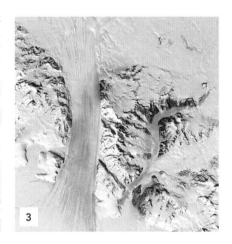

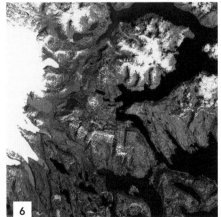

1. *(p.6)* British Antarctic Survey pyramid tent on Sibelius Glacier, Alexander Island. *(Photograph: Peter Fretwell 2018)*

2. *(p.9)* ESA Sentinel 2 image of Wilhemina Bay, western Antarctic Peninsula.

3. *(p.12)* Landsat 7 image of the Byrd Glacier cutting through the Transantarctic Mountains.

4. *(p.10-11)* Landsat 7 image of the Nordenskjöld Coast, north-east Antarctic Peninsula.

5. *(p.28)* Landsat7 image of Shirase Glacier flowing into the frozen Lutzow-Holm Bay, East Antarctica.

6. *(p.46)* ESA Sentinel 2 false colour image (using near infrared) of the central part of Kerguelen Island, showing the different colour of the lakes. The Red in this image denotes vegetation.

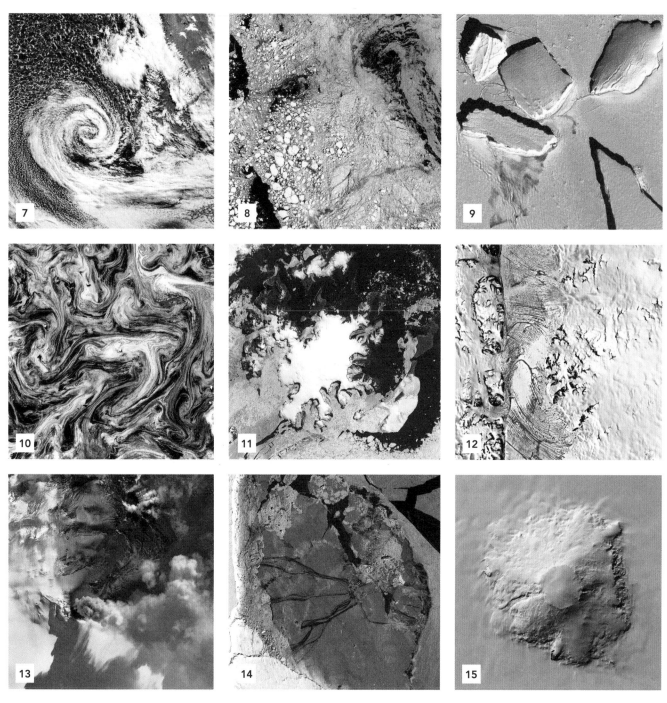

7. *(p.70)* MODIS satellite image of a large storm depression nearing the southern coast of Chile in the eastern Pacific.

10. *(p.122)* ASTER satellite image of sea ice near the South Orkney Islands. *(Image processing: Andrew Fleming)*

13. *(p.190)* Quickbird2 image of the eruption of Mount Belinda, Montagu Island, South Shetland Islands in 2006. *(Image: Maxar [Digital Globe])*

8. *(p.84)* EAS Sentinel 2 image of winter sea ice near the South Sandwich Islands. Saunders Island is just visible in the bottom left of the image.

11. *(p.150)* Landsat8 image of James Ross Island and other surrounding Islands on the north-east side of the Antarctic Peninsula.

14. *(p.192)* Emperor penguin tracks across newly formed sea ice, Worldview3 image , near the Halley Bay emperor penguin colony. *(Image: Maxar [Digital Globe])*

9. *(p.106)* Worldview3 high resolution image of an emperor penguin colony on the Riiser-Larsen Ice Shelf. *(Image: Maxar [Digital Globe])*

12. *(p.184)* Sentinel 2 image of meltwater pools on the George VI Ice Shelf.

15. *(p.202)* Sentinel2 image of Mount Takahe, Marie Byrd Land, West Antarctica.